Space, Place, and Landscape in Ancient Greek Literature and Culture

This book brings together a collection of original essays that engage with cultural geography and landscape studies to produce new ways of understanding place, space, and landscape in Greek literature from the fifth and fourth centuries BCE. The authors draw on an eclectic collection of contemporary approaches to bring the study of ancient Greek literature into dialogue with the burgeoning discussion of spatial theory in the humanities. The essays in this volume treat a variety of textual spaces, from the intimate to the expansive: the bedroom, ritual space, the law courts, theatrical space, the poetics of the city, and the landscape of war. And yet, all of the contributions are united by an interest in recuperating some of the many ways in which the ancient Greeks in the archaic and classical periods invested places with meaning and in how the representation of place links texts to social practices.

KATE GILHULY is Associate Professor of Classical Studies at Wellesley College. She is the author of *The Feminine Matrix of Sex and Gender in Ancient Greece* (Cambridge University Press, 2009). She began her research on cultural geography while a Fellow at the Radcliffe Institute for Advanced Study at Harvard University. She has published articles on the erotic geography of ancient Greece and is currently at work on a book on this topic.

NANCY WORMAN is Professor of Classics and Comparative Literature at Barnard College and Columbia University. She is the author of articles and books on style, performance, and the body in Greek culture and literature and is currently completing a book on landscape aesthetics in ancient literary theory and criticism. With Joy Connolly she is coediting the *Oxford Handbook of Ancient Literary Theory and Criticism*, and on her own she is devising a new project on "tragic bodies," which explores the aesthetics of embodiment in Greek tragedy and beyond.

Space, Place, and Landscape in Ancient Greek Literature and Culture

Edited by

Kate Gilhuly

Wellesley College

Nancy Worman

Barnard College

Cambridge University Press

CAMBRIDGE
UNIVERSITY PRESS

32 Avenue of the Americas, New York NY 10013-2473, USA

Cambridge University Press is part of the University of Cambridge.

It furthers the University's mission by disseminating knowledge in the pursuit of education, learning and research at the highest international levels of excellence.

www.cambridge.org
Information on this title: www.cambridge.org/9781107042124

First published 2014

A catalogue record for this publication is available from the British Library

Library of Congress Cataloguing in Publication data
Space, place, and landscape in ancient Greek literature and culture /
edited by Kate Gilhuly, Nancy Worman.
pages cm
Includes bibliographical references and index.
ISBN 978-1-107-04212-4 (hardback)
1. Space and time – Social aspects – Greece. 2. Landscapes – Social aspects – Greece. 3. Cultural geography – Greece. 4. Greece – Civilization – To 146 B.C. 5. Space in literature. I. Gilhuly, Kate, 1964– editor of compilation. II. Worman, Nancy, 1963– editor of compilation.
DF78.S69 2014
304.2′30938–dc23 2014002756

ISBN 978-1-107-04212-4 Hardback
ISBN 978-1-107-61853-4 Paperback

CONTENTS

CONTRIBUTORS

ALASTAIR J. L. BLANSHARD, Paul Eliadis Professor of Classics and Ancient History, the University of Queensland

CAROL DOUGHERTY, William R. Kenan Jr. Professor of Classical Studies, Wellesley College

CHRIS ECKERMAN, Assistant Professor of Classics, University of Oregon

KATE GILHULY, Associate Professor of Classical Studies, Wellesley College

ALEX PURVES, Associate Professor of Classics, University of California, Los Angeles

TIM ROOD, Dorothea Gray, Fellow and Tutor in Classics, St. Hugh's College, Oxford

NANCY WORMAN, Professor of Classics, Barnard College, Columbia University

ACKNOWLEDGMENTS

We are grateful to the institutions and grants that supported us during the time we were organizing the workshop that was the springboard for this volume. We formulated our first ideas for it while we were both on leave, Kate at the Radcliffe Institute for Advanced Study and Nancy visiting on a Loeb Fellowship at the Harvard Department of Classics. We applied for and received a Mellon SIRT Grant that, together with additional funding from a Barnard College Mini-Grant, paid for participants' attendance at the workshop, which was hosted at Wellesley College.

We also thank our contributors to the volume, who have been such lively and productive interlocutors in its development, as well as Betsey Robinson, who participated in the early stages of our discussions. We are grateful to the anonymous readers of the volume, who made many thoughtful and helpful suggestions for its improvement, and to our editor at Cambridge, Beatrice Rehl, who together with Isabella Vitti shepherded it through production with alacrity and precision.

On a personal note, Kate would like to thank Aidan and Anabel for providing her inspiration; Anne Gilhuly for her unstinting support and encouragement; and Mark Bush for his generosity of spirit. Kate dedicates her work on this volume to the memory of her father, Robert Thomas Gilhuly. Nancy would like to thank Iakovos Vasiliou, as always.

INTRODUCTION

Kate Gilhuly and Nancy Worman

Meaningful Vistas

This volume brings together a collection of essays that seek to recuperate some of the many ways in which the ancient Greeks in the archaic and classical period invested places with meaning, as well as how the representation of place in literature connects to social practices. While archeologists and ancient historians have long been attentive to geography in ancient Greece, in the past thirty years the innovative work of cultural geographers and scholars in related disciplines (e.g., social anthropology, landscape studies) has raised new possibilities for the examination of actual and symbolic topographies in ancient Greek culture. The essays gathered here explore the critical potential that a consideration of space, place, and landscape brings to the study of Greek literature and culture.

A number of ancient historians and archeologists have been profitably influenced by the "spatial turn,"[1] a term used to describe the confluence of interests across many disciplines regarding what it means to be situated in space. Scholars now conceive of the ancient world as comprised of cultural spaces that are not just the backdrop for the built environment but dynamic and multilayered social constructs whose possible

1 E.g., Horden and Purcell 2000; Van Dyck and Alcock 2003; Robinson 2011. Cf. Osborne 1987; Cole 2004.

meanings are created and negotiated in a variety of ways. Thus historians and archeologists have undertaken cross-cultural considerations of the development of ancient cities,[2] as well as exploration of cities or religious centers in their relationships to other places and regions, as these are made up of local ecologies and networks of communication and connection.[3] This means that scholars are considering defined types of space in terms of the wider cultural systems in which they are situated.[4] Landscape theory has been especially productive for thinking about the development of sacred space and its relationship to its broader context, as well as the intersection of landscape, monument, and memory.[5]

At the same time, theorists of classical literature are coming to grips with the importance of place and space for understanding textual strategies and cultural values. Since the 1980s, the powerful impact of structuralism on classical literary studies has played a central role in raising awareness of space and place in ancient texts. This conforms with the fact that space is central to the concerns of structuralism; as Doreen Massey has remarked, "The aim of structuralism in fact seemed to be to put space, rather than time on the intellectual agenda."[6] Indeed, spatial notions permeate the structuralist readings of

2 Smith 2003; Gates 2011.

3 Osborne 1987; Cole 2009. Scott 2013 considers the spatial dimension of the metropolis-colony relationship as it evolves over time, and applies spatial theory to a range of other historical questions, exploring the potential that an understanding of space has for scholars of Greek and Roman history. Horden and Purcell 2000 brings together broad-ranging historical essays on the relationship between people and their environment in the Mediterranean region.

4 Nevett 2010 collects and evaluates archeological evidence of domestic spaces to examine how they open up onto larger social and cultural issues. Cf. Nevett 1999, which argues for a recalibration of our understanding of gender in the context of domestic space. O'Sullivan 2011 explores the way Romans thought about moving through society in terms of power, the body, and space in an exploration of walking in Rome.

5 Scott 2010; Cole 2004; Alcock and Osborne 1994; Alcock 2002.

6 Massey 2005: 36.

Jean-Pierre Vernant and Pierre Vidal-Naquet, as well as others these groundbreaking scholars of ancient literature have influenced. Consider Vernant's analysis of the binary of space and movement as reflected in the conjunction of Hestia and Hermes on the base of the pedestal of Zeus' statue at Olympia.[7] In his *Origins of Greek Thought,* when describing the move away from the Mycenaean Palace system toward the *polis,* Vernant presents this transformation as having a central spatial dimension:

> The terms a Greek would use to express this concept are striking: he would say that certain deliberations, certain decisions must be brought *es to koinon* [to the commons], that the ancient privileges of the king and *arche* itself were set down *es to meson,* in the middle, at the center. The recourse to a spatial image to express the self-awareness that a human group has acquired, its sense of existing as a political unit, is of value not only as a comparison; it also reflects the creation of a social space that was altogether new.[8]

In the first section of the *Black Hunter*, entitled "Space and Time," Vidal-Naquet sees a similar transition in spatial coordinates from Homer to the classical period: "As presented by the *Odyssey*, space figures into the opposition between the real and the imaginary, the gods, monsters, and men, sacrifice and barbarism. After Homer it became the city-state's space. . . . "[9]

While many of the authors in this collection have been deeply influenced by structuralist approaches, the theoretical work in cultural geography and landscape studies, much of which also reflects a poststructuralist awareness of place and landscape as contested and constructed, has had a more immediate impact on most of us, as we explore further below. Our volume seeks to converse with and extend the engagement of other scholars of classical literature and

7 Vernant 1963.

8 Vernant 1962: 47.

9 Vidal-Naquet 1986: xxii.

culture who also incorporate a similarly evolving mix of concepts and methodologies.[10]

With this collection of papers we offer a distinctive supplement to this discussion that both encompasses a number of different authors and genres and takes up from different angles contemporary questions about place, space, and landscape. In the modern period, these three key terms have shifted repeatedly and cultural geographers use them to mean different things;[11] Ancient Greek terminology seems just as unstable as modern geographical discourse, since there is considerable overlap and fluidity in the usage of such terms as *topos* ("place, site"), *chôros* ("space, region"), and *chôra* ("land, countryside"). In what follows, we make a brief attempt to separate out the modern terms in ways that are particularly useful for and relevant to the ancient Greek context, as well as specific to the focus of the papers in this volume. It should be noted, however, that the contributors to this volume use these terms in a variety of ways, not only because of their unique emphases but also because of their exposure to changing awareness in the disciplines about the ways in which modern terminology encodes ideological concerns.[12]

Let us consider, as a first instance, the concept of place. Theorists who engage with it as a social construct, as opposed to a strictly real, inhabited place or undifferentiated space, have proved most useful to those of us who focus on the confluence of particular settings and issues of status and identity. This includes an awareness of perspective, since the looker effectively shapes the place.[13] Consideration of place makes visible relationships among topographical features, identity,

10 E.g., Romm 1992; Collot et al. 1997; Thalmann 2011.

11 Kenneth Olwig (2002a, 2002b) notes the influence of the capacious German concept of *Raum*, which can mean space, place, and landscape, and has led to considerable fluidity in the separating out of "space" from "place" and both of these from "landscape."

12 Olwig (2002a, 2002b) has identified such shifts in terminology as the result of ideological pressures.

13 See, e.g., Calame 2007.

and the body, while exploration of the vantage from which place is considered reveals the way scenes can be shaped by ideological perspectives and manipulated to promote authority and stature.[14]

As David Harvey has argued, a place is a multilayered locus of the imaginary, a site for the negotiation of social relations, a form of discourse and of power, and, at the same time, a material practice.[15] The construction of place is also relational: take, for example, the ways in which the city of Thebes functions in Athenian drama. As Froma Zeitlin has famously argued, it serves as the other to Athenian values and identity.[16] A place can be defined in its difference relative to other places that are constantly evolving, "woven together out of ongoing stories," as Doreen Massey says.[17] For scholars who study the past primarily through texts, a focus on place is tantalizing because it insists that the imaginary is grounded in the real, locating the discursive patterns that we interrogate in the realm of lived experience. It thus provides the potential to understand how and where the imaginary interacts with actual places and practices.

The philosopher J. E. Malpas observes that the social and place are mutually constitutive: "the social does not exist prior to place nor is it given expression except in and through place.... It is within the structure of place that the very possibility of the social arises."[18] This perspective prompts us to consider the particularity of place – what are the social, political, and economic relationships that imbue a place

14 See Chris Eckerman's essay in this volume, "Pindar's Delphi" (Chapter 1) for an exploration of terms and consideration of the way the framing of place operates in the construction of an authoritative identity.

15 Harvey 1996: 293–94.

16 Zeitlin 1990.

17 Massey 2005: 126. Kate Gilhuly's essay in this volume, "Corinth, Courtesans, and the Politics of Place" (Chapter 5), looks at the identity of Corinth in relationship to Athens as an ongoing process.

18 Malpas 1999: 36. Cf. Cresswell 2004: 33, who suggests that we think of place as a "'necessary social construction' – something we have to construct in order to be human."

with a distinctive character? What are the local attributes of a place that give it an identity? How can this identity be manipulated?

Politically minded theorists urge that places are not static and bounded but are the site of movement, process, and practice. For the Hellenist, chronologically sundered from the subject of study, captive to fragmentary and seemingly unchanging representations of Greek culture, the allure of place lies in its porousness, the suggestion of plenitude, and the relation of place to the practices of everyday life. As Arturo Escobar writes, we must recognize "that place, body, and environment integrate with each other; that places gather things, thoughts and memories in particular configurations; and that place, more an event than a thing, is characterized by openness rather than by a unitary self-identity."[19]

If place and the social are crucially interdependent, and if places are performative accretions of local meanings, it follows that identity and its coordinates (e.g., class, gender, sexuality) are implicated in the construction of place, and these are constantly resurfacing emphases in this collection of papers. Depictions of place encode a range of values – political, ethical, and aesthetic – as they are embedded in actual practices. A focus on place thus has the potential to foster an approach to an embodied kind of meaning, the space of lived experience. Such insights can lead, for instance, to recognizing more fully the fluid, performative character of urban settings such as the Athenian agora. Drama and oratory help to give the marketplace shape both through actual ritual and legal practices and in the often derisive attributes they attach to it in the course of arguments and plotting. An exploration of place in Greek literature should, then, add dimensions to our understanding of identity in the ancient world.

By contrast to place, the notion of space contributes a different vantage to this exploration. Insofar as the theoretical vocabulary suits the ancient phenomena, "space" would

[19] Escobar 2001: 143; cf. de Certeau 1984: 117, who claims this event status for space. See further below.

appear to be the most abstract, encompassing, and fluid designation: "what begins as undifferentiated space becomes place as we get to know it ... "[20] While some theorists emphasize the opposition between space and place, others note the fluidity of the relationship between the two.[21] Andrew Merrifield understands the interaction between these two terms as a constant negotiation; he argues that space and place "are different aspects of a unity," and the distinction between the two exists "only insofar as it represents different 'moments' of a contradictory and conflictual process."[22]

Space may oppose inside and outside, include symbolic and actual registers, and encompass areas as disparate as those defined by dramatic and topographical coordinates. And yet, as the geographer Kirsten Simonsen has recently emphasized, one great contribution of Henri Lefebvre's influential work *The Production of Space* is his focus on space as embodied, as just the opposite of abstract or formal.[23] Lefebvre regards the abstraction of space, its disengagement from the warm life of the body, as a result of the visualizing, signifying work of drawing generalized meanings from specific, lived engagements. But paradoxically it is the body itself, what Lefebvre calls the "spatial body," that generates the coordinates demarcating space in abstract terms (e.g., left and right, up and down).[24]

The spatial body in literature can orient the reader to the text. One gets a keen sense of the body's spatial orientations on the Homeric battlefield, for instance, where the reader's high

20 Tuan 1977: 6.

21 In "The Permeable Spaces of the Athenian Law-Court" (Chapter 7) Alastair Blanshard finds flexibility in the ways in which Athenians understood and designated the spaces of law. He argues that law-courts were performed places, thus showing the fluidity between space and place in the disposition of the Athenian Agora.

22 Merrifield 1993: 527. See also Harvey 1996: 315, who also describes the relationship between space and place as dialectic.

23 Simonsen 2005; cf. Lefebvre (1974) 1991: 194–96; also Rodaway 1994.

24 Cf. Tuan 1975: 215; also Cook 1998, who refers us to Aristotle's *Physics* 4.1.208b.

prospect (essentially a gods-eye view) allows for the placed viewing of spears whipping through the air from left and right, cutting across and through warrior bodies, which take the fall from high up in the chariot to down on the dusty ground.[25] This same spatial body may also give shape to intimate spaces, as its close-range prospects correlate most directly to domestic settings and thus to indoor and outdoor, floor and ceiling, as points of reference.[26] In tension with such abstractions are the phenomenological features of bodies in spaces – that is, what bodies do in detail, when engaged in particular practices in particular spaces, within the social sphere or in private.[27]

There is a point at which notions of space may include specific places, especially in the context of self-consciously mimetic places like the theatre.[28] The notion of representational space in general is important for our volume since we are engaging spatiality through texts. Representational space allows for a distinctive type of multilayering. For performative space can, at times, engage simultaneously with real and fictional places. In this spatial play, there is a convergence of real and imaginary, and these different experiences of space and/or place are imbricated.[29]

Some scholars have argued that Lefebvre effectively inverts more common notions of space and place, so that the former sometimes appears to indicate specific places.[30] Thus, for instance, while discussing the spatial organization of ancient Athens, he includes the Agora and Acropolis under this rubric

25 See Purves 2006.

26 Cook 1998.

27 Taking a close look at such details, Alex Purves works within this small-scope notion of space in her essay "In the Bedroom: Interior Spaces in Herodotus' *Histories*" (Chapter 3), where she traces important parallels between entering and exiting a room and the putting on and taking off of clothing.

28 See, e.g., Wiles 1999; Worman 2001; Rehm 2002.

29 Nancy Worman's essay "Mapping Literary Styles in Aristophanes' *Frogs*" (Chapter 6) directly addresses the interplay between mimetic and real landscapes of Attica.

30 E.g., Cresswell 2004: 10.

and yet refers to the agora as "gathering-place."[31] But some of the ways in which the terms have shifted over the past decades may also obscure where they intersect, and why Lefebvre and other poststructuralist thinkers such as Michel de Certeau, writing in the 1970s and early 1980s, may have seen space where their critics, writing in 1990s and after, saw place.[32] With the rising influence of historicizing narratives and identity politics, the specifics of place emerged as more compelling sites for scrutinizing power and its negotiations than the relatively abstract schemes that notions of space tend to illuminate.[33] And yet, as the essays in this volume demonstrate, spatial coordinates may bring into sharper focus aspects of status and identity, while characteristics associated with a place may have little to do with a specific site.

In any case, for some of us in this volume it is useful that theorists such as Lefebvre sustain a focus on the continuum from small-scale and embodied coordinates to larger and more abstract ones.[34] In considering ancient phenomena, we are constantly reminded that space can also delineate vast expanses, high vantage points with long views across zones that may run from the upper air to below the ground. From early on Greek poets and philosophers strove to conceive of the earth's edges and to organize Greek and foreign identities in relation to them.[35] From this broad prospect space, in coordination with time, may entail movement within (often quite expansive) confines. Unlike the notion of place, however, space does not necessarily concern so much individual identities

31 Lefebvre (1974) 1991: 249–50.

32 Cf. de Certeau's 1984: 117 emphasis on space as "actuated by the ensemble of movements deployed within it" and as a "practiced place."

33 In fact, Cresswell's 2004 introduction to his essay, which is very grounded in particular histories and locales, appears to claim for place what earlier writers might have taken to fall under rubrics such as space or landscape.

34 See esp. chapter 4 in Lefebvre (1974) 1991; also Tuan 1977.

35 Bakhtin (1930s) 1981; Lefebvre (1974) 1991: 237–41; Cook 1998: 558–59; cf. Romm 1992.

as ontological categories in a broad and symbolic sense (e.g., borderlands, which tend to be inhabited by monsters and other hybrid figures).[36]

Again, place seems to be much more specific than this, a distinct site of marked convergence that accrues social and cultural details over time. Thus place is also historical, existing in narrative time.[37] It has a past that can give a sense of rootedness, as well as one of loss and thus nostalgia.[38] These aspects usually situate the individual in relation to both actual and typifying coordinates. Our third term, "landscape" also implies a subject, one who surveys a geographical vista. This term is freighted with historical, cultural, and especially aesthetic meaning. When, for instance, Denis Cosgrove seeks an example of how topographies are shaped into value-laden patterns by the perceiver, he looks to Ruskin, a profoundly influential interpreter of European art and architecture in the nineteenth century. Cosgrove remarks, "[L]andscape became in Ruskin's eyes a suitable subject for examining the deepest moral and artistic truths," describing a sensibility steeped in centuries-old traditions of landscape painting and architecture.[39]

Landscape embraces aspects of layered cultural settings as they are familiarly viewed, such as the Athenian theater and temple precincts, as well as internationally celebrated sanctuaries such as Delphi and Olympia. Scholars have thus come to recognize that notions of landscape may include aspects more akin to the ways in which place and space intersect with embodied perspectives and social orientations. In the introduction to his influential edited volume, *Landscape and Power*, W. J. T. Mitchell remarks, "Landscape as a cultural medium . . .

36 E.g., Homer, *Od.* Bks. 9–10; Hesiod, *Theog.* 211–332; Euripes, *Her.* 359–424.

37 Carol Dougherty's essay, "Ships, Walls, Men: Classical Athens and the Poetics of Infrastructure" (Chapter 4), explores the shifting significance of Athenian terms for infrastructure during the last third of the fifth century BCE.

38 On *lieux de memoire* see Nora 1989; Schama 1999; Tuan 2003.

39 Cosgrove and Daniels (1988) 2002: 5. Cf. Cosgrove 1985, 1998; also Tuan 1995.

has a double role with respect to something like ideology: it naturalizes a cultural and social construction, representing an artificial world as if it were simply given and inevitable, and it also makes that representation operational by interpellating its beholder in some more or less determinate relation to its givenness as sight and site."[40] Landscape from this perspective has less to do with the history of aesthetics and representation and more with cultural negotiation, with issues similar to the kind emphasized by theorists focused on place. As Mitchell maintains, one must address the ways in which "landscape *circulates* as a medium of exchange, a site of visual appropriation, a focus for the formation of identity."[41] Such conceptualizations suggest contention and struggle, not only (or not so much) for the land itself as for a way of seeing it that carries enough force as to appear to be reality rather than interpretation.[42]

Olwig has also put forth a sense of landscape as an inhabited setting forged by the intersection of the natural lay of the land and human negotiations relating to power and the law.[43] Landscape from this perspective is always interactive and substantive, less abstracted in symbolic, aesthetic, or discursive terms than enacted in and lorded over. As such it intersects more fully with notions of place and space, but with the distinction that topography and scenery serve as meaningful coordinates in navigating social status, ownership, and civic or national identity.

What, then, can we claim about such awareness in classical antiquity, seemingly long before traditions were established and conventions became so ingrained as to inform what one saw in the lay of the land? Does it make sense to say, for

40 Mitchell (1994) 2002: 2.

41 Mitchell (1994) 2002: 2.

42 Tim Rood's essay, "Space and Landscape in Xenophon's *Anabasis*" (Chapter 2), works this out on a large and yet concertedly human scale that emphasizes perspective, considering how the Persian expanse shaped by Xenophon forges a distinct patterning of the viewed landscape as a space that captures ideological differences between Greeks and Persians.

43 Olwig 2002b.

example, that in Pindar's odes the poet-performer celebrates ritual centers such as Olympia and Delphi as, like Ruskin, a received landscape? Are ancient landscapes in fact just as infused with traditions that organize associations of place and cultural meaning? As the papers in this volume demonstrate, we think that they are.

All of the authors in this collection have been influenced to varying degrees by landscape studies and cultural geography, but this collection does not advance any one particular theoretical stance or present a comprehensive view of these fields. The foregoing discussion of terminology is meant only to provide an aerial view, and each writer invokes and defines geographical terms specifically for his or her argument. Rather, we hope to initiate a discussion of the intricate relationships among space, place, landscape, and ancient Greek texts. Texts are like places in that they can contain multiple voices and perspectives, they mediate between practice and ideology, and they are multiply interpretable. The language we use to describe features of literature – plot, topos, point of view – suggests an overlapping of literature with space, place, and landscape. We think that these essays demonstrate how literary texts participate in the construction of place and illuminate the nexus between experience and ideology and, conversely, that they show how cultural geography is useful for understanding the space of literature.

Further, we aim with this approach to bring together elements of representation that disorient a traditional understanding of familiar texts by posing questions that open out onto lived realities in particular places and geographies that were familiar to ancient audiences but have not always been so apparent to modern readers. For this reason, the focus on place or space itself presents interpretive challenges that the texts may not appear to present in and of themselves. Paying attention to the human dimension of the lay of the land provokes questions about the relationship between representations and the material conditions in which they were produced, as well as the extent to which representations – and the ideological perspectives that inform them – themselves foster "realities."

While our focus is on the classical period in Greece and engages with literary, historical, and rhetorical texts, these essays also speak to larger concerns that have ramifications for other areas within the discipline – a discipline that is after all fundamentally defined by geography. And insofar as specific angles on larger topics may help to redefine them, we hope that this volume will expand awareness of the poetics of place, space, and landscape in the study of the ancient world.

THE PAPERS IN THIS VOLUME

We have noticed certain intersections between and among the contributions in this volume and have arranged them to showcase these resonances. We begin with Chris Eckerman's essay, because his work engages intimately with some of the perspectives articulated above. In addition, Pindar's poetry is especially evocative of place, and his manipulation of landscape deftly illustrates the relationships among topography, identity, and the body. In "Pindar's Delphi," Eckerman discusses the ways in which Pindar's poetry appropriates and manipulates the resonances of Delphi. Complicating the traditional interpretation that Pindar's deictic evocation of place indicates that these poems were performed at Delphi, he suggests that Pindar's poetry encouraged audiences to engage in vicarious transport to that sacred landscape. He draws on Yi-Fu Tuan's observation that space is constructed in relation to the body, exploring the ways in which Pindar recruits the senses to promote a vivid and visceral experience of Delphi from afar. Geographical features such as the Kastalian Spring were implicated in various types of identity formation and could be manipulated for their ideological power. Eckerman makes the important observation that the experience of place is historically contingent, pursuing in addition the ramifications of the ancient belief that certain features of the earth were divinities, such as rivers and springs. Thus the landscape was conceived of as animated, and gender and power dynamics were implicated in one's relationship to place. Further, narratives associated with Delphi lent prestige to the landscape and to those whom Pindar associated with it.

The next two papers map diametrically opposed visual perspectives on space and/or landscape in historical texts. Tim Rood takes an expansive view in "Space and Landscape in Xenophon's Anabasis," considering the march to the sea and along the Black Sea coast. Rood suggests that through his depiction of space and landscape, Xenophon constructs an idealized version of leadership and a sense of Greek cultural identity. For instance, he notes that the Greeks' encounter with Persian space is characterized by the language of encirclement and aporia ("no way out"). The image of encirclement resonates with the Greek perception of their encounters with Persians as "a combat between Greek virtue and Persian numbers." The desperate position they find themselves in, as conveyed by the recurring language of aporia, depicts their ultimate passage as evidence of Greek resourcefulness. In describing his own military valor, Xenophon represents himself as able to see more than other generals and thus also able to lead his troops to safety. The alien quality of the land draws in greater relief the cultural identification of Greek soldiers. Rood's paper shows how power is constructed through the mastery of space and landscape.

Where Rood treats vast outdoor expanses, Purves takes us inside to focus on intimate space. With "In the Bedroom: Interior Spaces in Herodotus' *Histories*," Purves turns from a view out of the window (i.e., landscape) to those through the bedroom door (i.e., bounded space). She highlights the narratives in which Herodotus emphasizes entering and exiting the private spaces that consolidate kingship and thus power. In exploring these dynamics, Purves urges attention to the ways in which the self is implicated within private spaces, most especially because of the intimate acts that occur there, including dressing and undressing. She thus also traces important parallels in the *Histories* between entering and exiting a room and the putting on and taking off of clothing, emphasizing how the fraught alliance between intimacy and power is repeatedly orchestrated around the body and its close confines. This also entails that as these narratives move in and out of such spaces, looking and desiring are centrally intertwined. Purves reveals how Herodotus plots the reference points framing interior spaces in such a way as to reveal their nature as a trap, which entails that there can be no way out except through violence or mastery of some other sort.

The next two chapters engage with Athens' civic imaginary, exploring how dominant discourses projected and reflected Athens' self-image. In "Ships, Walls, Men: Classical Athens and the Poetics of Infrastructure," Carol Dougherty charts the evolution of Athenian ideas about what defines the city as they are reflected in the famous account of the wooden walls oracle in Herodotus' *Histories*, in Sophocles' *Oedipus Tyrannos*, and in Thucydides' *History of the Peloponnesian War*. She identifies ships, walls, and men as key terms for Athens' self-image as it rises to the height of its power and confronts the brutal consequences of its imperial overreach. In an analysis that seeks out the locus where the lived experience of the city converges with discourse, Dougherty engages with Athens as it explores what truly defines a city – is it the local environment, its mastery of vast expanses, or its inhabitants? In this chapter, we see the changing valence of these terms: in Herodotus' *Historie*s the wooden walls of the ship are the city's salvation. In the *Oedipus*, the imagery of which Dougherty reads as a means of interrogating Athenian choices, Sophocles depicts ships and walls as inextricably bound up with one another, as models for the Athenian city, while the inhabitants are the animating force. Thucydides emphasizes instead the ambiguity in the imagery of ships and walls: do they protect or confine? He also ironizes the terms of Athens' self-definition – ships and walls – and then traces their ultimate annihilation, along with the men who inhabited them.

Kate Gilhuly takes us beyond infrastructure to contrasts between cities and their reputations. In "Corinth, Courtesans, and the Politics of Place," she explores the way Athens constructed its own image through its imagining of Corinth.

Gilhuly thus emphasizes the relational aspects of place, how a place cannot exist in isolation and is shaped in contrast to other entities. Focusing on the comic stereotype – that all Corinthians were involved in prostitution – she correlates Athenian depictions of Corinth with discourses surrounding the sex trade. She argues that, because prostitution had become a category that encapsulated a range of ideological distinctions, this characterization of Corinthians could serve as a prism through which Corinthian difference was interpreted, in terms of politics, economics, culture, and geography. The courtesan is deployed to comment on Corinth's oligarchic constitution, its permissive trade environment, and its status as a gateway to the East and a crossroads to tourists and traders on land and on sea. She suggests that since the courtesan is a figure of identification as well as difference, this characterization also refracts Athenian anxieties about its own status as it replaced Corinth as the most powerful port city in Greece.

The next chapter initiates a turn toward mimetic space. In "Mapping Literary Judgment in Aristophanes' *Frogs*," Nancy Worman draws attention to the literary critical coordinates of the path that Aristophanes charts from the city through the countryside and into the underworld. The Athenian landscape and the route out of the city, as these are represented onstage, correspond to modes of literary style and have ethical undertones. Worman examines the way the changing topography depicted in the comedy gives a spatial component to the differences between the tragic poets, as Dionysus passes through it to the space of literary judgment. Thus the comic scheme implicitly associates Euripides with the marsh and the creative and unconventional music of the frogs, while it characterizes Aeschylus by the more traditional lyric *topos* and its denizen: the bee in the meadowland. Further, the passage along the banks of the Ilissos reinforces poetic comparisons using water imagery. Where Euripides is linked to the poetics of water as sparkling and fluid, the excesses of Aeschylus' "natural" style are associated with the flood. But comedy also evaluates tragedy by means of this mediating journey that exposes the flaws of both poets, revealing Euripides as too urbane and Aeschylus too weighty and wild.

Alastair Blanshard shifts our attention to the place of judgment as a performed space that intersects with citizen identity in distinctive ways. In "The Permeable Spaces of the Athenian Law-Court," Blanshard uses textual and archeological evidence to argue that the court was constituted within various places by means of legal rituals and accoutrements. It was also a space in which the typical coordinates of citizen identity were singularly challenged. Despite archeologists' assumptions and willful identification of remains in the Agora, there seems to have been no single, central building that served as a permanent court, and Blanshard argues that multi-use open-air buildings could be rendered legal spaces through practices that constituted a space as a place of law. This helps to explain, for instance, why the lampooning in Attic comedy around setting up a court any old place by means of moveable objects (i.e., paraphernalia pertaining to court practices) would have played so familiarly for Athenians. As Blanshard emphasizes, it is by means of the ritualized use of such items that a sense of place and the establishment of occasion are achieved. And yet, as he also details, the juror's day in court is an occasion like no other, since formal procedures were put in place to insure that citizen jurors could not consort with their friends and allies. Thus the isolation of the juror within the ritualized legal space, in combination with orators' warnings about tricks of argument and subterfuge, served to inhibit his sense of self as a citizen well placed in a familiar world and known to his peers.

In our disposition of these essays, things are gained and lost. We ourselves have noticed other intriguing correlations between and among the papers. In fact, we see a number of important relationships emerging in the pieces that work in fluid combinations, such as body and place, landscape and identity, ritual and space, and discourse and topographies. Further, the individual pieces of these categories may work in different combinations (e.g., place and identity, landscape and the body). Many of our papers embrace a fuller set of these components, or take the concepts themselves in different directions. Above all, by refusing to craft a reductive model that synthesizes all the conceptions of place, space, and landscape that applies to all the essays collected here, we hope to highlight the multiple ways these geographical ideas can work together in productive ways to represent the literary spaces, places, and landscapes of ancient Greece.

Bibliography

Alcock, S. 2002. *Archeologies of the Greek Past: Landscape, Monuments, and Memories*. Cambridge.

Alcock, S., and R. Osborne. 1994. *Placing the Gods: Sanctuary and Sacred Space in Ancient Greece*. Cambridge.

Bakhtin, M. M. (1930s) 1981. *The Dialogic Imagination: Four Essays*. ed. M. Holquist, tr. C. Emerson and M. Holquist. Austin, TX and London.

Calame, C. 2007. "Longus and the Pastoral Garden of Eros." In *Sacred Gardens and Landscapes: Ritual and Agency*, ed. M. Conan. Washington, DC. 43–54.

Cole, S. G. 2004. *Landscapes, Gender, and Ritual Space: The Ancient Greek Experience*. Berkeley.

Cole S. G. 2009. "I Know the Number of the Sand and the Measure of the Sea" in *Geography and Difference in the Early Greek World*, ed. K. Raaflaub and R. Talbert. Oxford. 197–214.

Collot, M. et al. 1997. *Les enjeux du paysage*. Bruxelles.

Cook, A. 1998. "Space and Culture." *New Literary History* 29.3: 551–72.

Cosgrove, D. 1985. "Prospect, Perspective and the Evolution of the Landscape Idea." *Transactions of the Institute of British Geographers* n.s. 10.1: 45–62.

1998. *Social Formation and Symbolic Landscape*. 2 ed. Madison, WI.

Cosgrove, D. and S. Daniels. (1988) 2002. *The Iconography of Landscape: Essays on the Symbolic Representation, Design and Use of Past Environments*. Cambridge.

Cresswell, T. 2004. *Place: A Short Introduction*. Oxford.

de Certeau, M. 1984. *The Practice of Everyday Life*, trans. S. Rendell. Berkeley. (Orig. Paris 1980.)

Escobar, A. 2001. "Culture Sits in Places: Reflections on Globalism and Subaltern Strategies of Localization." *Political Geography* 20: 139–174.

Gates, C. 2011. *Ancient Cities: the Archaeology of Urban Life in the Ancient Near East and Egypt, Greece and Rome*. London.

Harvey, D. 1996. *Justice, Nature and the Geography of Difference*. Oxford.

Horden, P. and N. Purcell. 2000. *The Corrupting Sea: A Study of Mediterranean History*. London.

Lefebvre, H. (1974) 1991. *The Production of Space*, tr. D. Nicholson-Smith. Oxford

Malpas, J. E. 1999. *Place and Experience: A Philosophical Topography*. New York.

Massey, D. 2005. *For Space*. London.

Merrifield, A. 1993. "Place and Space: A Lebfvrian Reconciliation." *Transactions of the Institute of British Geographers* N.S. 18.4: 516–530.

Mitchell, W. J. T. ed. (1994) 2002. *Landscape and Power*, 2nd ed. Chicago.

Nevett, L. 1999. *House and Society in the Ancient Greek World*. Cambridge.

2010. *Domestic Space in Classical Antiquity*. Cambridge.

Nora, P. 1989. "Between Memory and History: *Les Lieux de Mémoire*." *Representations* 26:7–24.

Olwig, K. 2002a. "The Duplicity of Space: German *Raum* and Swedish *Rum* in English Language Geographical Discourse." *Geographiska Annaler, Series B: Human Geography* 84.1: 1–17.

2002b. *Landscape, Nature and the Body Politic: From Britain's Renaissance to America's New World*. Madison, WI.

Osborne, R. 1987. *Classical Landscape with Figures: The Ancient Greek City and Its Countryside*. Dobbs Ferry, NY.

O'Sullivan, T. 2011. *Walking in Roman Culture*. Cambridge.

Purves, A. 2006. "Falling Into Time in Homer's 'Iliad.'" *CA* 25.1: 179–209.

Rehm, R. 2002. *The Play of Space: Spatial Transformation in Greek Tragedy*. Princeton.

Robinson, B. 2011. *Histories of Peirene: A Corinthian Fountain in Three Millennia*. Princeton.

Rodaway, P. 1994. *Sensuous Geographies: Body, Sense, and Place*. London.

Romm, J. 1992. *The Edges of the Earth in Ancient Thought: Geography, Exploration, and Fiction*. Princeton.

Schama, S. 1999. *Landscape and Memory*. New York.

Scott, M. 2010. *Delphi and Olympia: the Spatial Politics of Panhellenism in the Archaic and Classical Periods*. Cambridge.

2013. *Space and Society in the Greek and Roman Worlds*. Cambridge.

Simonsen, K. 2005. "Bodies, Sensations, Space and Time: The Contributions of Henri Lefebvre." *Geografiska Annaler, Series B: Human Geography* 87.1: 1–14.

Smith, M. 2003. *The Social Construction of Ancient Cities*. Washington, DC.

Thalmann, W. G. 2011. *Apollonius Rhodes and the Spaces of Hellenism*. Oxford.

Tuan, Y.-F. 1975. "Space and Place: Humanistic Perspective." *Progress in Geography*, 6: 211–52.

1977. *Space and Place*. London.

Tuan, Y.-F. 1995. *Passing Strange and Wonderful: Aesthetics, Nature and Culture*. Washington DC.

2003. "Perceptual and Cultural Geography: A Commentary." *Annals of the Association of American Geographers* 93.4: 878–81.

Van Dyck, R. M. and S. E. Alcock. 2003. *Archaeologies of Memory*. Malden, MA.

Vernant, J.-P. 1962. *Origins of Greek Thought*. New York.

1963. "Hestia-Hermès: Sur l'expression religieuse de l'espace et du mouvement chez les Grecs." *L'Homme* 3.3: 12–50.

Vidal-Naquet, P. 1986. *The Black Hunter: Forms of Thought and Forms of Society in the Greek World*, Trans. A. Szegedy-Maszak. Baltimore and London.

Wiles, D. 1999. *Tragedy in Athens: Performance Space and Theatrical Meaning*. Oxford.

Worman, N. 2001. "The *Herkos Achaiôn* Transformed: Character Type and Spatial Meaning in the *Ajax*." *CP* 96: 230–54.

Zeitlin, F. I. 1990. "Thebes: Theater of Self and Society in Athenian Drama." In *Nothing to Do with Dionysos? Athenian Drama in Its Social Context*, ed. J. Winkler and F. I. Zeitlin. Princeton. 130–67.

1

PINDAR'S DELPHI

Chris Eckerman

What should we call Delphi? A place? A space? A landscape? More specifically, what should we call Apollo's sanctuary at Delphi, as Pindar depicts it in his epinician odes? Place, space, and landscape comprise the holy trinity of contemporary spatial thinking, and with the metaphor of the trinity comes the image of the interfusion of all three phenomena. I suggest that we need all three terms to think through Delphi in Pindar's poetry. Examining place, space, and landscape in Pindar's odes, then, this chapter addresses many issues related to contemporary geographic studies, such as the tension between real and imagined landscapes, the relationship between human bodies and land, the cognitive associations between places and their human-given names, and the role of place in relation to power and identity. By focusing on the role of Delphi in Pindar's odes, I further the interests of contemporary Pindaric scholars who root epinician discourse in its sociohistorical context.[1] Epinician poetry's geographic rootedness, however, has not received extended critical interest, and here I suggest that a greater sensitivity to geography opens Pindaric criticism in productive directions, by pointing out how Pindar constructs narrative to allow his audience to experience Delphi's space,

1 For example, Kurke 1991; Dougherty 1993; Nicholson 2005.

regardless of performance place, and how he manipulates Delphi's sacredness to suit his patrons' ideological needs.

In this chapter I analyze space, landscape, and place through individual case studies. The first section examines the role of the body (and the senses particularly) as the vehicle through which space is experienced in *Pythian* 6. The second part, focusing on landscape, raises epistemological and ideological questions in relation to Pindar's use of Delphi's landscape-divinity Kastalia. In the third section, on place, I consider how place is constructed through myth and how place is marked by humans to develop connections to it. For example, Pindar composes etiological myth for the temple of Apollo in his eighth *Paean*, and he uses Apollo's temple to reaffirm his patron's elite status in *Pythian* 7. In closing this section, I look at Pindar's evocation of Arkesilas IV's dedication of charioteering equipment at Delphi, and show how Pindar integrates the dedication into *Pythian* 5 to reaffirm his patron's power. One theme dominates these three sections on space, landscape, and place: Pindar's appropriation and manipulation of Delphi's symbolic capital in his own interest as well as in the interest of his patrons.

This chapter positions itself among the burgeoning secondary literature that stresses the ideological importance of Panhellenic sanctuaries to the colonial elite, which comprised a considerable portion of Pindar's patrons.[2] Several of the odes this chapter addresses in some detail were composed for colonists, and when Pindar constructs Delphic space, narrates a myth about Delphi, or mentions one of its monuments for his colonial patrons, they connect to Delphi and thereby reaffirm their Hellenicity, whether as patron, family member of the victor, or member of the general Greek audience. As Malkin remarks, "One should not be surprised that the list of victors in the ancient Olympic and Pythian Games included so many Greeks from the west. They wanted that Greekness and enhanced the importance of the notion

2 See, with reference to further bibliography, Scott 2010; Neer 2007; Morgan 1990.

while practicing it."[3] Panhellenic sanctuaries, such as Delphi, provided venues for aristocratic conspicuous consumption, and an examination of Pindar's engagement with Delphi provides a further glimpse into the negotiations of power and identity that were worked out at Panhellenic sanctuaries in the archaic and classical periods.

CONSTRUCTING SPACE THROUGH COGNITION AND THE SENSES: THE PROEM OF PYTHIAN 6

Pythian 6's proem provides Pindar's most extensive depiction of Delphi in the epinician odes. Pindar constructs a procession that moves to the "enshrined navel" (ὀμφαλὸν ... νάϊον, ll.3–4) of the earth, where a "treasury of hymns" (ὕμνων θησαυρὸς, ll. 7–8) has been built for Xenokrates, who won the prestigious chariot victory at the Pythian games of 490 BC.[4] Xenokrates, an Emmenid, was the brother of Theron and a member of Akragas' elite.[5] The ode references and praises Xenokrates' son, Thrasyboulos, more than Xenokrates, however, and this may suggest that Thrasyboulos commissioned the ode on behalf of his father's victory. After developing the proem, which focuses on Delphic imagery, Pindar creates structural similarities between Thrasyboulos and Antilochus, who, by dying for his father, Nestor, exhibited exemplary filial piety.[6] After the first two stanzas, then, Pindar redirects his attention, focusing on topics that do not include Delphi. Accordingly, I address only the first two stanzas here:

Ἀκούσατ᾽· ἦ γὰρ ἑλικώπιδος Ἀφροδίτας ST. A
ἄρουραν ἢ Χαρίτων
ἀναπολίζομεν, ὀμφαλὸν ἐριβρόμου
χθονὸς ἐς νάϊον προσοιχόμενοι·

3 Malkin 2003: 71.

4 On the date, see Morrison 2007b: 41.

5 Theron seems to have become tyrant in 489. For discussion, see Morrison 2007b: 42.

6 Gildersleeve 1885: 314 says that "the glory of filial devotion" is the theme of the ode.

Πυθιόνικος ἔνθ' ὀλβίοισιν Ἐμμενίδαις
ποταμίᾳ τ' Ἀκράγαντι καὶ μὰν Ξενοκράτει
ἑτοῖμος ὕμνων
θησαυρὸς ἐν πολυχρύσῳ
Ἀπολλωνίᾳ τετείχισται νάπᾳ·

τὸν οὔτε χειμέριος ὄμβρος, ἐπακτὸς ἐλθών ST. B
ἐριβρόμου νεφέλας
στρατὸς ἀμείλιχος, οὔτ' ἄνεμος ἐς μυχούς
ἁλὸς ἄξοισι παμφόρῳ χεράδει
τυπτόμενον. φάει δὲ πρόσωπον ἐν καθαρῷ
πατρὶ τεῷ, Θρασύβουλε, κοινάν τε γενεᾷ
λόγοισι θνατῶν
εὔδοξον ἅρματι νίκαν
Κρισαίαις ἐνὶ πτυχαῖς ἀπαγγελεῖ.

Listen! for indeed we are plowing up the field of dark-eyed Aphrodite and of the Graces, while we proceed to the enshrined navel of the loudly-rumbling land, where for the prosperous Emmenidai and for rivery Akragas, and especially for Xenokrates, a Pythian victor's ready treasury of hymns has been built in Apollo's valley rich in gold:

one which neither winter rain, coming brought in from a loudly-rumbling cloud, like a relentless army, nor wind shall lead into the depths of the sea, beaten by the irresistible silt. But in clear light its facade will proclaim a chariot victory, famous in the words of men, shared by your father, Thrasyboulos, and your clan, in the dells of Krisa.[7]

What does Pindar's procession to Delphi, as well as his evocation of Delphic phenomena, clarify regarding the performance place of this ode? Epinician critics nearly unanimously assume that *Pythian* 6 was first performed at Delphi,[8] but, as argued

7 Text and translations are based on those of Race 1997.

8 Scholars who read the passage as evidence of Delphic first performance include: Athanassaki 2011: 254–55; Neer 2001: 287; Clay 1999: 33; Race 1997: 312; Bernardini et al. 1995: 541; Gelzer 1985:

elsewhere, the ode most likely would not have been presented at Delphi because epinician odes were generally first performed in the home *poleis* of the victors and because the necessities of commissioning *Pythian* 6 make it unlikely that the ode could have been enacted at the Pythian festival before it was over.[9] In an important article on epinician poetry, moreover, Nancy Felson has noted that "one device, in the vast arsenal of poetic devices, is the powerful linguistic tool of deixis, which can engage the emotions of persons listening to any sort of text and give them the illusion of participating in events and places and times that are far distant."[10] To use Felson's term, Pindar is able to "vicariously transport" his audiences through spaces and to places that he references in his odes.[11] The ode, then, was most likely first performed in the patron's home polis, Akragas.

Pindar's construction of Delphi in this proem fits remarkably well with the sentiments of one contemporary geographic theorist's idea of space. Yi-Fu Tuan notes, "If geometrical space is a relatively late and sophisticated cultural construct, what is the nature of man's original impact with his world, his original space? ... We can say little more than that original space possesses structure and orientation by virtue of the presence of the human body. Body implicates space; space coexists with the sentient body ... visual perception, touch, movement, and thought combine to give us our characteristic sense of space."[12] Tuan's work emphasizes the importance of the body in relation to lived space, but his work is good to think with in relation to Pindaric poetry precisely because Pindar constructs space in narrative in such a way that it feels real for the audience.

98–99; Burton 1962: 15 and 17; Farnell 1932: 183; Wilamowitz 1922: 139; Bury 1892: 29–30; Boehmer 1891: 3; Boeckh (1821) 1963 *ad loc.*

9 On the performance of *Pythian* 6 particularly, see Morrison 2007b: 43; Eckerman 2011a: 1–8. On the performance of epinician poetry at Panhellenic sanctuaries generally, see Eckerman 2012.

10 Felson 1999: 1.

11 Morrison 2007b: 81 too refers to vicarious transport to Delphi upon reperformance of *Pythian* 6 in a non-Delphic space.

12 Tuan 1975: 215.

For the perception of space, Tuan emphasizes the importance of bodies and their senses, and his description of space aligns with Pindar's poetic tactics in *P.* 6, since, as I argue, Pindar integrates visual perception, sound, movement, and touch in his proem to help his audience experience Delphic space.[13] With a phenomenological approach to space in this section, then, I suggest that Pindar's proem uses the senses, imagination, and kinesthesia to encourage vicarious transport to Delphi.

The poem opens with a striking imperative, listen (ἀκούσατ', l.1), and Pindar thereby stresses one of the five senses, hearing. Assuming that Pindar here uses the imperative in his ode to recreate the victory ceremony at which a herald would have announced Xenokrates to be the victor in the chariot competition, Gildersleeve explains the imperative as a herald's cry.[14] But, as Acosta-Hughes notes, it is not uncommon for Greek poems to open with ἀκούσατ' or a similar imperative, because "the use of direct second person address is a typical feature of all poetry that evokes the image of a public audience, including epigram, didactic epic, and others."[15] Pindar's ode did have an audience, and, although the performance of all odes before their audiences entails a visual and aural connection between audience and chorus, the imperative brings the chorus and audience particularly close together, since the chorus thereby demand's the audience's attention.[16] Moreover, the imperative makes the audience active participants in the performance of praise, and, as the audience members participate in the performance of the ode by viewing and listening, they reaffirm the ode's ideology and are thereby implicated in the celebration of Xenokrates and the powerful Emmenid clan at Akragas.

13 On the importance of the body for the production of space in this volume, see Gilhuly's and Worman's Introduction (7–8); and Purves's discussion in this volume (Chapter 3: 96–99).

14 Gildersleeve 1885: 316.

15 Acosta-Hughes 2002: 37.

16 On the solo vs. choral debate on the performance of epinician poetry, see, most recently, Eckerman 2011b.

Immediately after demanding the audience's attention, Pindar, introducing another sense, uses language that emphasizes touch and kinesthesia. The haptic connection that Pindar first develops through the image of plowing a field (ἄρουραν . . . ἀναπολίζομεν, ll. 2–3) fades away as the audience realizes that the chorus plows a metaphorical field, that is, constructs poetry.[17] Nonetheless, the image of plowing may have been chosen to relate to the choreography of the ode, because, on occasion, Greek dances mimicked agricultural acts.[18] If *Pythian* 6's choreography included dance moves that mimicked plowing, the dance creates a further tactile relationship with the land, as the chorus mimics cutting into the land (ἄρουραν, l.2) with their plows.[19]

Pindar thereafter encourages his chorus and audience members to envision themselves processing to the temple of Apollo, where the *omphalos* was housed, and touch and kinesthesia are further developed. When the chorus says that they are moving to "the enshrined navel (*omphalos*) of the loudly-rumbling earth" (ll.3–4),[20] Pindar leaves it open to the audience to interpret whether they are moving from the place of performance to Delphi or are already located at Delphi and are processing along the Sacred Way to the temple of Apollo. Regardless, as Hetherington remarks in an article on the geographies of touch

17 See too *O.* 9.26; *N.* 6.32; *N.* 10.26 and Bernardini et al. 1995: 540.

18 See, for example, Xenophon, *Anabasis* (6.1) on the *karpaia* dance, "some Aenianians and Magnesians stood up and began a dance in armour called the *karpaia*, which goes like this: one man puts down his weapons and starts to sow grain and drive a team, while constantly turning this way and that as though through fear; a robber approaches and the farmer spots him, grabs his weapons, goes to meet him, and fights him to stop him stealing his team of oxen. They keep time throughout with the music of the pipes" (trans. R. Waterfield). We know next to nothing about epinician choreography; cf. Carey 2007: 199.

19 This is speculative, since we have lost the choreography for Pindar's odes.

20 Following Herman's conjecture νάϊον for the codices' ναόν. As Farnell 1932: 184 notes, this passage should refer to the *omphalos* enshrined within the temple of Apollo.

and placemaking, "Touch is a way of removing doubt – of confirming."[21] Pindar's audience, then, "imaginatively confirms" their presence at Delphi by processing to Apollo's temple.

The ode's verbal and participial morphology, moreover, encourages the audience to make themselves subjects in the ode and to transport themselves to Delphi. As Morrison writes, the plural verb and participle forms allow the audience to weave themselves into the imagined movement.[22] The audience hears "we are plowing" (ἀναπολίζομεν, l.3), and they thereby become a part of the inclusive "we" of the first-person plural subject, and they continue to position themselves as subjects in the participle "processing" (προσοιχόμενοι, l.4), which comes shortly after the verb. As Heath notes, moreover, *P.* 6 and *N.* 2 are the only Pindaric odes that do not contain first-person singular morphological forms.[23] Pindar, accordingly, does something unusual here by keeping the performing subject in the plural, and the grammar of the ode too encourages vicarious transport.

At the same time that he develops a haptic connection to Delphi with processing, Pindar introduces another word, "loud-rumbling" (ἐριβρόμου, l.3), which, like the first word of the ode, "listen" (ἀκούσατ'), stresses the importance of the aural zone. Pindar does not clarify how he wants his audience to interpret the adjective "loud-rumbling," but, as others note, the adjective likely refers to earthquakes.[24] By striking the earth with their feet, furthermore, the chorus may have created a "loud-rumbling" earth during performance, since, as Hardie remarks, "Choric activity comprised song and dance (the sound of the mouth as well as that of the feet)."[25] The

21 Hetherington 2003: 1941.

22 Morrison 2007a: 114.

23 Heath 1988: 192.

24 See Bernardini et al. 1995: 542. We should not follow Gildersleeve 1885: 316–17, who assumes that the adjective refers to the waters of the spring-nymph Kastalia, since "loud-rumbling" would not be an ideal way to describe the sound of water from a spring.

25 Hardie 1996: 224.

audience would thereby hear the aural phenomena that are characteristic of Delphi's "loud-rumbling" landscape. Pindar elsewhere integrates aural terminology that resonates with the choreography of the ode in just such a manner.[26]

After bringing the audience to Delphi through imagining a procession and having the audience experience Delphi's loud rumbling, Pindar introduces the image of a treasury, which would be familiar to his audience given the numerous treasuries that were built at Delphi.[27] Pindar plays with his audience, since they first imagine the material treasuries at Delphi. Pindar's treasury, however, is a nonmaterial "treasury of hymns" (ὕμνων θησαυρὸς, ll. 7–8), and the metaphoric treasury of hymns refers to the ode itself, a hymn. Moreover, Pindar modifies his treasury with the adjective "ready" (ἑτοῖμος), which may programmatically introduce the superiority of Pindar's treasury over material treasuries. Since Pindar stresses that his treasury is "ready" and currently being performed, the audience may infer that a material treasury, on the other hand, is "not ready" for Xenokrates and accordingly cannot rival Pindar's treasury. The adjective, however, most obviously stresses the immediate performance of the ode.

In the first stanza, then, through sound, touch, and visualization, the audience feels itself to be at Delphi, while, in the second stanza, Pindar furthers an embodied experience of Delphi by having the audience envision and vicariously experience meteorological phenomena. Pindar, accordingly, intensifies the audience's sensory experience of Delphi when he introduces winter rain (χειμέριος ὄμβρος, l. 10) and wind

26 See Hardie 1996: 222. In the sixth *Paean*, the chorus refers to beating the ground with their feet: ἦτορι δὲ φίλῳ παῖς ἅτε ματέρι κεδνᾷ/ πειθόμενος κατέβαν στεφάνων/ καὶ θαλιᾶν τροφὸν ἄλσος Ἀ/πόλλωνος, τόθι Λατοίδαν/ θαμινὰ Δελφῶν κόραι/ χθονὸς ὀμφαλὸν παρὰ σκιάεντα μελπ̣[ό]μεναι/ ποδὶ κροτέο[ντι γᾶν θο]ῷ (for in heeding my own heart, as a child obeys his dear mother, I have come to Apollo's precinct, nurse of crowns and feasts, where the maidens of Delphi often sing to Leto's son at the shady navel of the earth and beat the ground with a rapid foot," trans. Race, ll. 12–18).

27 On Delphi's treasuries, see Partida 2000.

(ἄνεμος, l.12). The simile of the rain as a relentless army (στρατὸς ἀμείλιχος, l.12) further stresses the inescapable damage that will befall physical treasuries at Delphi. Furthermore, the rain's arrival from a "loud-rumbling cloud" (ἐριβρόμου νεφέλας, l. 11) is striking, since Pindar here uses the adjective "loud-rumbling" (ἐριβρόμου) a second time, when he could have chosen another descriptive adjective for the sake of variation. The intensification of the Delphic scene through the addition of dangerous meteorological phenomena encourages a particularly vivid transport to Delphi, but Pindar's audience is safe because it only experiences the ravage of the elements vicariously through the ode.

Pindar first heightens his audience's experience of Delphi as they imagine the possibility that Delphi might be hit by a wind so severe that it could roll the material treasuries of Delphi into the depths of the sea, that is, into the Corinthian gulf (ἐς μυχοὺς ἁλὸς, ll. 12–13). The treasuries at Delphi, on the southern edge of Mt. Parnassos, and the Corinthian gulf were far removed from one another by the plain of Krisa, and no storm could send a treasury into the sea. Pindar's hyperbole is effective, then, as he creates a storm so powerful that it could do the impossible and role the treasuries at Delphi into the Corinthian gulf. Even such a powerful storm could never damage Pindar's "treasury of hymns," however, because, as an ode, it is impervious to the ravages of inclement weather. Pindar thereby develops his hierarchical relationship between material and literary culture, suggesting that his superior art form does not suffer the ravages of the elements.[28]

After the imagined rain and wind, Pindar strikingly shifts the image. He envisions pure light (φάει ... ἐν καθαρῷ, l.14) as a backdrop, uses δὲ (l.14) to note the transition to his new topic (his treasury of hymns), and refers to his treasury's façade (πρόσωπον, l.14). As argued elsewhere, Pindar encourages his audience here to imagine his treasury's façade as being

28 See Horace *Odes* 3.30 for further development of what would become a literary *topos*.

decorated with an equestrian scene, "a victory, glorious with a chariot" (εὔδοξον ἅρματι νίκαν), perhaps a sculpted frieze or pedimental sculpture.[29] As part of the treasury's decoration, the audience may imagine something akin to the eastern pediment of the Alkmaeonid temple of Apollo at Delphi (c. 510 BC), which included a sculpted image of Apollo in a chariot.[30] Sculptures such as the famed Delphi charioteer too provide analogous epinician programs. Accordingly, although Pindar's treasury is a nonmaterial treasury, it derives its symbolic decoration from its relation to historical treasuries and prominent equestrian sculptural programs at Delphi.

In *Olympian* 6, dated nearly two decades after *Pythian* 6, Pindar also uses architectural imagery to describe the construction of an ode. In the opening of the ode, he analogizes the front of a building to the beginning of an ode. There the façade (πρόσωπον) of a building must be visible from afar (τηλαυγές, l.4), just as the beginning (πρόσωπον) of an ode should be conspicuous. *Olympian* 6 and *Pythian* 6, then, both use monumental architecture, the πρόσωπον of a public building specifically, as a vehicle to understand poetic structure. By mentioning Xenokrates at line 6 in the first stanza, Pindar places Xenokrates in the beginning (πρόσωπον) of his ode, while he also encourages his audience to visualize Xenokrates' treasury as being decorated with an equestrian scene on its façade (πρόσωπον). Pindar uses synaesthesia as the audience both listens to the *prosopon* of Pindar's ode and visualizes the *prosopon* of Xenokrates' treasury, and such double reference has been recognized elsewhere in Pindar's corpus.[31]

There is a notable discrepancy in verb tenses, since Pindar states that his treasury "has been built" (τετείχισται, l.9) in Apollo's valley rich in gold (i.e., Delphi) and also claims that its façade "will announce" (ἀπαγγελεῖ, l.18) an equestrian victory. Pindar first positions his treasury in the physical

29 For discussion, see Eckerman 2011a.

30 Bommelaer 1991: 53–55, 182.

31 See, for example, Hardie 1996: n. 50.

landscape at Delphi with the perfect indicative because his metaphorical treasury needs to be positioned in Delphi's landscape to develop the associations of a traditional treasury built at Delphi. However, the use of the future indicative (ἀπαγγελεῖ) is characteristic of Pindar's use of other future indicatives, which, as Bundy first noted, refer to the immediate performance of the respective ode.[32] The treasury's façade "will" and "does" announce the victory of Xenokrates, then, with the enunciation of the verb (ἀπαγγελεῖ) during performance, and the perfect passive indicative verb, "has been built" (τετείχισται), does not provide evidence for the performance location of this ode, although it does stress how important the landscape of Delphi is to Pindar's ode.[33]

When read together, Pindar's treasury and its façade position Xenokrates and the Emmenids more among their fellow aristocrats than among their fellow citizens. Pindar asserts that his "treasury of hymns" (ὕμνων θησαυρὸς) has been built for the fortunate Emmenidai, for Akragas, and for Xenokrates (ὀλβίοισιν Ἐμμενίδαις ποταμίᾳ τ' Ἀκράγαντι καὶ μὰν Ξενοκράτει, l.6). With the inclusion of the clan name, Emmenidai, and the personal name (accompanied by the emphatic particle μάν), Pindar stresses the importance of the individual's accomplishment in relation to his family, and Xenokrates' position as an aristocrat who has a treasury "built" at Delphi encourages the audience to recollect the dedications of individual tyrants at Delphi.[34] Pindar, however, popularizes the treasury by asserting that it has also been built for "rivery Akragas." Akragas' broader citizenry, then, shares in the victory, and Pindar skillfully places Xenokrates both among his fellow aristocrats as well as among his fellow citizens. The treasury's façade

32 Bundy 1962: 21. On such Pindaric futures, see particularly Slater 1969.

33 Morrison 2007b: 42 too notes that τετείχισται need not refer to a performance context at Delphi.

34 For a discussion of the dedication of treasuries by tyrants at Panhellenic sanctuaries, see, with reference to further bibliography, Neer 2007: 243–45.

(πρόσωπον), on the other hand, announces a victory that is glorious specifically for Thrasyboulos' father and for the Emmenid clan (πατρὶ τεῷ, Θρασύβουλε, κοινάν τε γενεᾷ, l. 15), presumably because the façade is decorated with Xenokrates' victory. Pindar does not assert that the façade will be glorious for Akragas. As Harrell has noted,[35] colonial elites negotiated their position in relation to their fellow aristocrats and in relation to their fellow citizens with their performances and dedications at Panhellenic sanctuaries, and Pindar here, although his discourse is somewhat inclusive, positions Xenokrates and the Emmenids more squarely among their fellow elites than among their fellow citizens.[36]

The ode's monostrophic meter has led some scholars to suggest that the ode was performed during a procession,[37] but I would argue that the monostrophic meter creates an atmosphere of procession that helps the audience vicariously experience a procession while the ode is performed statically. As Morrison notes, "the procession of the opening has become metaphorical by vv. 7–9, where Pindar announces 'a ready store-house of songs.'"[38] Pindar and Bacchylides composed processional odes (προσόδια), but it seems unlikely that any of Pindar's or Bacchylides' epinician odes would have been composed for procession, since an epinician chorus wants to keep the audience's attention and should not risk losing the audience by processing during the ode.[39] The monostrophic meter of *Pythian* 6, then, does not provide evidence that *Pythian* 6

35 Harrell 2002.

36 In the ode's primary myth, Pindar constructs analogues between the Emmenid, Thrasyboulos, and the hero Antilochus, thereby also positioning the Emmenids among their fellow elite forerunners.

37 Burton 1962: 15, 24; Gelzer 1985: 98–99.

38 Morrison 2007b: 42.

39 As Rutherford 2001: 715 notes, the term προσόδια is problematic: despite a name that would suggest processional performance for the genre, "in the fragments of ancient προσόδια, how much evidence is there for processional performance? Comparatively little, in fact."

was performed processionally,[40] but, since monostrophic odes were occasionally processional, Pindar's use of the monostrophic meter in *Pythian* 6 adds to the atmosphere of procession that plays a fundamental role in the proem.[41]

Perhaps the most notable aspect of Pindar's construction of Delphic space in this passage is the manner in which Pindar teases participation out of his audience through their bodies to help them experience Delphi. Pindar requires his audience's attention from the first word of the poem and, once he has it, he moves his audience to Delphi with processing, while referencing the rumblings of Delphi's sacred earth and thereby providing both a tactile and aural connection to Delphi. After stressing the harm that the material treasuries suffer at the hands of the elements, Pindar focuses attention on his "treasury of hymns," and the audience visualizes treasuries at Delphi. One might have expected Pindar to detach from the landscape of Delphi, since it suffers the elements, when he introduces his ontologically superior poetry, but the image of the treasury reasserts the importance of Delphi to Pindar's poetics and to his patron, Xenokrates, whose Pythian victory Pindar celebrates in this ode. As Felson notes, "Deictics lend poetry a dynamic quality. Not only do they attract attention and contribute vividness (a sense of visual presence), but they situate audiences in space and time. Indeed, no one hearing them can easily refrain from imaginatively traveling along the pathways which they designate: they contribute to a poetic enchantment, or θέλξις, that is impossible to resist."[42]

40 *P.* 12, also for an Akragantine victor, is monostrophic; see Morrison 2007a: 114.

41 As Rutherford 2001: 719 notes, there was a category of *prosodion* that was "vicariously processional," in which a chorus performed statically, while a procession of sacrificial victims was led to the altar. Although the case of *prosodia* that Rutherford refers to and the case of *Pythian* 6, as I argue here, are "vicariously processional" in different ways, the vicarious *prosodion* is a helpful analogue because it reminds us that hymns that were connected to procession were regularly performed statically.

42 Felson 1999: 8–9.

It seems likely, then, that *Pythian* 6 was not composed for first performance at Delphi, yet the ode creates Delphic space at every performance. As Felson further notes:

> Pindar uses deixis with expertise and subtlety, primarily to make his audiences "travel" across space and time. First he locates them in the text either at the site of victory or at the victor's hometown. Then he transports them along carefully demarcated pathways, ultimately returning them (again, in the text) to their place of origin. They experience vicarious travel through a vividness, or illusion of seeing (*enargeia*), that he linguistically creates.[43]

I have emphasized phenomenological aspects of space in this section to show how space is experienced through cognition and the senses, both real and imagined. Pindar, then, integrates the place of victory, Delphi, into the space of performance for both rhetorical and ideological reasons, and the ode's audiences and readers will respond to Delphic space as they see fit, regardless of the performance place of this ode.[44]

LIVING WITH AN ANIMATE LANDSCAPE: A CASE STUDY WITH KASTALIA

Contemporary geographers emphasize the manufactured character of landscapes, stressing that landscapes are sections of land that are perceived and constructed by human beings.[45] Yet a starkly different concept of landscape exists between contemporary geographers and many ancient Greeks because contemporary geographers generally assume that the surface of the earth is inanimate, while mainstream Greek religious

43 Felson 1999: 5.

44 In this volume (Chapter 7: 262), Blanshard similarly discusses how Aeschines, in *Against Ctesiphon*, takes the jury to the *Stoa Poikile* in their imagination.

45 For an overview and reference to further bibliography, see Wylie 2007: 17–93.

Weltanschauung animated the surface of the earth. Greek spring nymphs, such as Delphi's Kastalia, are a case in point. Kastalia complicates modern Western notions of landscape by forcing us to consider human interaction with Kastalia not in terms of animate beings (humans) and an inanimate object (water), but rather in terms of human beings and a divinity.

Nymphs, such as Kastalia, play an important role within Pindar's odes, in which he regularly fashions them as figureheads of their *poleis*. As Larson notes, "In the odes of Pindar . . . an individual nymph is elevated to represent the city itself; she personifies at once the land, its familiar topographical features, and the local mythic genealogy. These nymphs take honored positions at the beginning or end of many odes and are usually apostrophized, or addressed directly, by the poet."[46] In Pindar's odes, Kastalia plays a prominent role within her cultural matrix, and here I show how Pindar situates her within her geographical and cultural context and how he uses her in the interest of his patrons.[47]

Pindar displays a substantial knowledge of Delphi's landscape, and Kastalia serves as a good example for Pindar's familiarity with Delphi's topographic details. In the previous section on *Pythian* 6, I stressed the importance of sound and visualization in Pindar's poetics, and Pindar integrates sound and visualization into his poetic program in relation to Kastalia also. In the sixth *Paean*, for example, the narrator describes a procession at Delphi and refers to bronze-gated Kastalia being bereft of dancing at lines 7–9:

ὕδατι γὰρ ἐπὶ χαλκοπύλῳ
ψόφον ἀιὼν Κασταλίας
ὀρφανὸν ἀνδρῶν χορεύσιος ἦλθον. . . .

For, having heard the sound of Kastalia, at the bronze-gated water, being bereft of the dancing of men, I have come. . . .

46 Larson 2001: 37.

47 Pindar evokes Kastalia seven times in his epinicians and once in his paeans (*O.* 7.17, 9.17; *P.* 1.39, 4.163, 5.31; *N.* 6.37, 11.24; *Pae.*6.8). Bacchylides mentions her once (B. 3, 20).

With the epithet "bronze-gated" (χαλκοπύλῳ, l. 7), Pindar refers to Kastalia's archaic fountain house, which was located about 500 meters east of the *temenos* of Apollo;[48] as the scholiast notes, the river-god Kephisos flowed into Kastalia through the mouths of bronze lions.[49] The ode seems to have been first performed at Delphi, and Pindar's first audience would experience Delphi's space themselves. Yet, by making an accurate reference to Kastalia with "bronze-gated," Pindar gains authority through his knowledge of specific on-site details, and the adjective encourages later audience members and readers to visualize Kastalia's bronze gates, regardless of re-performance venue.

The narrator, presumably to be construed with Pindar, claims that he came to Delphi because he could "hear" (ἀιὼν, l. 8) the waters of Kastalia and, at the same time, he could hear the waters of Kastalia being "bereft" (ὀρφανὸν, l.9) of the choral dancing of young men. Ironically, the narrator "hears" Kastalia's waters, but he does not "hear" what he thinks he should hear, namely the sound of a male chorus performing beside Kastalia. The narrator claims that the absence of the sound of choral dancing motivated his travel to Delphi to correct the situation. With this rhetorical trope, Pindar creates a narrator who has supernatural powers, able to hear Kastalia from far away, say Thebes, and who, at the same time, plays the role of a savior, coming to Delphi and reconciling Kastalia's unfortunate situation through the immediate performance of this paean by a chorus of males.[50]

48 See Hardie 1996: 219.

49 On the archeology of Kastalia, see Amandry 1977, 1978. On the archaic fountain, see Amandry 1977: 198–225.

50 Hardie 1996: 220 would take the passage further, "The sequence of ideas is made more complex by an ambiguity in the word order. If ὕδατι is taken with Κασταλίας, and ψόφον with ἀνδρῶν, another construction is available: "for having heard, at the bronze-gated water of Castalia, the noise – bereft of males – of dancing ... " On this version, ψόφον would itself represent the noise of choric dancing, and not of Castalia. That more than one valid syntactical construction is available is unlikely to be accidental, and double reference may be involved."

In addition to creating a vivid landscape for Delphi by evoking Kastalia's visual and aural qualities, Pindar aligns Kastalia with the ideological interests of his patrons. Scholars regularly refer to Kastalia as a periphrasis for Delphi in Pindar's odes, but, as Hardie has rightfully stressed, this is an impoverishing way to refer to Kastalia. He writes,

> Parke noted that Pindar, like other poets, can use the name as a convenient way of denoting Delphi and its environs. [Parke] suggested that ... it helps to build up local colour in a Delphian paean. This suggestion is obviously true, so far as it goes ... but ... it seems unlikely that Castalia is brought in simply as a feature of the local geography. A more complex field of reference, both sacral and musical, seems to lie behind Pindar's introduction of the fountain.[51]

Following Hardie, I would suggest that we should not refer to Kastalia as a periphrasis for Delphi, because it divests Kastalia of her unique individuality and her sociocultural importance. I would add, however, that along with Kastalia's "sacral" importance comes, most importantly, Pindar's ability to manipulate it.

Examining passages in Pindar's odes where Kastalia has traditionally been referred to as a periphrasis for Delphi, we see that Kastalia fulfills a richer function than the term periphrasis denotes. For example, at *Olympian* 7.13–17 Pindar refers to victories that Diagoras of Rhodes won "beside Alpheios" (παρ' Ἀλφειῷ, l.17) and "beside Kastalia" (παρὰ Κασταλίᾳ, l.17):

> καί νυν ὑπ' ἀμφοτέρων σὺν Διαγόρᾳ κατέβαν, τὰν ποντίαν/ ὑμνέων, παῖδ' Ἀφροδίτας Ἀελίοιό τε νύμφαν, Ῥόδον,/ εὐθυμάχαν ὄφρα πελώριον ἄνδρα παρ' Ἀλφειῷ στεφανωσάμενον/ αἰνέσω πυγμᾶς ἄποινα/ καὶ παρὰ Κασταλίᾳ, πατέρα τε Δαμάγητον ἁδόντα Δίκᾳ.

> And now to the accompaniment of both, I have disembarked with Diagoras, singing a hymn to Rhodes of

[51] Hardie 1996: 221.

> the sea, the child of Aphrodite and bride of Helios, so that I may praise, in recompense for his boxing, that straight-fighting man of prodigious power, who won by Alpheios and by Kastalia, and may praise his father, Damagetos, who is favored by Justice.

One could say that Alpheios and Kastalia here serve as periphrases for Olympia and Delphi, but such a gloss would ignore the important parallelism that Pindar develops through Olympia's and Delphi's landscape divinities. As Lily Kong has noted, "if sacredness is not inherent, attention must be paid to how place is sacralized."[52] The evocation of Alpheios and Kastalia reminds the audience of the divinities that inhabit the landscapes of Olympia and Delphi, and, because most Greeks believed that landscape divinities such as Alpheios and Kastalia took active interests in human lives and affected the outcomes of human events, the integration of Alpheios and Kastalia into epinician odes aligns them in the interest of Pindar's patrons. Similarly, Pindar addresses Kastalia at *Olympian* 9.17 and thereby constructs himself as one who has the goddess on his side, since she presumably hears Pindar's apostrophe (9.14–18):

> αἰνήσαις ἓ καὶ υἱόν,/ ἃν Θέμις θυγάτηρ τέ οἱ σώτειρα λέλογχεν/ μεγαλόδοξος Εὐνομία. θάλλει δ' ἀρεταῖσιν/ σόν τε, Κασταλία, πάρα/ Ἀλφεοῦ τε ῥέεθρον·

> Praise the son and his city, which Themis and her glorious daughter, saving Order, have as their allotment. It flourishes with achievements by your stream, Kastalia, and that of Alpheios.

Larson has suggested that the nymphs play a non-elite role in Pindar's odes: "They seem to represent the polis itself or the land it sits on, as opposed to the more narrowly defined, elite groups who claimed descent from the heroic ancestor. The eponymous nymph of a city, she who is represented on its coins, cannot be claimed by any one family."[53] Larson is right to stress that not

52 Kong 2001: 213.

53 Larson 2001: 39.

only the elite can appropriate the nymphs, but I would argue that the integration of the Nymphs into epinician odes places the victor into cosmic space and in relation to the gods rather than in relation to his non-elite social classes. In terms of theodicy, moreover, Pindar's evocation of Kastalia aligns the will of Kastalia with the will of his patrons, and, in this regard, his integration of Kastalia mirrors his broader use of gods, whom he regularly fabricates as being in support of his patrons.

Since places are important for identity (personal, regional, national, ethnic, and so forth), people regularly associate themselves with prestigious places. Much as the myth of Alpheios and Arethusa connects mainland Greece with Magna Graecia, a myth connects Kastalia with the people of Lilaia in Phokis via the river-god Kephisos. The Lilaians asserted that Kephisos was the source of Kastalia at Delphi,[54] and, in the past, scholars would have referred to such geographic ties in terms of center and periphery, with the less important, "peripheral" site (such as Lilaia) using myth to connect to the more important, "center" site (such as Delphi).[55] More recent scholarship has encouraged us to avoid hierarchical terminology and to look at such relationships in terms of networks and rhizomes.[56]

Parke suggests that the reason for the Lilaian connection to Delphi was to rationalize the particularly copious and constant flow of Kastalia,[57] but, as Larson notes, "the desire to forge connections with Delphi and the famous spring probably explains the Lilaian story."[58] By manipulating their connection

54 Lilaia was on the northeastern slope of Mt. Parnassos. Pausanias says that the sound of Kephisos rising from the ground is like that of a bellowing bull (10.33.5).

55 Pausanias reports the physical connection between the river-god Kephisos and the spring-nymph Kastalia: "The strongest confirmation of this view is a custom of the Lilaeans, who on certain specified days throw into the spring of Kephisos cakes of the district and other things ordained by use, and it is said that these reappear in Kastalia" (10.8.5).

56 See, for example, Malkin 2003, 2011.

57 Parke 1978: 290.

58 Larson 2001: 148.

to their epichoric river-god, Kephisos, the Lilaians linked their comparatively insignificant, marginal *polis* with Apollo's politically and culturally significant sanctuary at Delphi. The story of Alpheios and Arethusa and the story of Kephisos and Kastalia are examples of the human imagination using geography to increase the importance of less prestigious sites through their relation to more prestigious sites.[59] Just as colonial Greeks in Sicily could connect to mainland Greece via Alpheios and just as Lilaians could connect to Delphi via Kephisos, so too all Greeks can connect to Kastalia and Delphi's prestigious landscape, when Pindar evokes her in a poem.[60]

The integration of Kastalia into epinician poems is powerful, since there is perhaps no better way to evoke Delphi's sacredness than through a divinity embodied in Delphi's land, and this takes us back to concepts of landscape with which I began this section. As members of an ethnic group, Greeks would have had some similar responses to Kastalia, stimulated by a shared literary, geographic, and sociocultural heritage, but nonetheless differences in personal and regional identity meant that individuals would also have had varying associations with Kastalia. For Greeks living nearby, Kastalia was a divinity to whom one should devote ritual. For Greeks who won victories at the Pythian games, Kastalia was an epichoric divinity that enabled their victories. For Theocritus and Roman poets, Kastalia was closely associated with poetic inspiration. For some, however, Kastalia might not be much more than a source of freshwater. Kastalia, accordingly, is a complex deity and landscape marker who means many things dependent upon how one receives, focalizes, and concomitantly "reads" her.[61]

59 Similarly, according to a story preserved in Strabo (6.2.4), if one threw a cup into Alpheios on the mainland of Greece it reappeared in Arethusa on Sicily.

60 On landscape and identity see Rood in this volume (Chapter 2: 89–92).

61 Athenian drama asserts, moreover, that the servants of the Delphic sanctuary washed their hair in Kastalia before entering the temple

With Kastalia, Pindar creates emotive vignettes of Delphi's landscape and he binds her ideologically to his poems of praise. As Duncan and Duncan note, "If landscapes are texts which are read, interpreted according to an ingrained cultural framework of interpretation, if they are often read 'inattentively' at a practical or nondiscursive level, then they may be inculcating their readers with a set of notions about how the society is organized: and the readers may be largely unaware of this."[62] Accordingly, numerous geographers argue that landscape is an "ideological veil" that allows the powerful to take control of the land.[63] Cosgrove notes, for example, that landscape is fundamentally a cipher used by those in power to legitimate their control,[64] and this description of the elite's manipulation of the land suits Pindar perfectly well. Pindar was a complicit agent in the reaffirmation of his patrons' importance in relation to Greek society and in relation to the cosmos, and epinician poetry was a vehicle that allowed Pindar to use Kastalia and the landscape of Delphi as he saw fit.

The Power of Delphi as a Place

Place in theorized discourse is not a synonym for location, because location does not include the personal, the historical, the mythical, and the communal aspects associated with a location that turn that location into a place. As Philip

of Apollo and that Kastalia's water was used for purifying the temple (Eur. *Ion* 144, Eur. *Ion* 94 ff., *Phoen*. 222.). Furthermore, visitors to the sanctuary, as an act of ritual ablution, sprinkled themselves with the water of Kastalia before entering the sanctuary (Heliodorus *Aithiop*. II 26). Theocritus and Roman poets developed an associative connection between Kastalia and poetic inspiration, no doubt thanks in part to the important role that Kastalia plays in Pindar's poetics (For discussion, see Parke 1978: 204–09).

62 Duncan and Duncan 1988: 123.

63 For an overview, see Wylie 2007, chapter 4.

64 "Landscapes sustain and elaborate the symbolic code of bourgeois society" (Cosgrove 1998: 10).

Sheldrake asserts, "the concept of place refers not simply to geographical location but to a dialectical relationship between environment and human narrative,"[65] and, as Ed Relph further notes, "place and sense of place do not lend themselves to scientific analysis for they are inextricably bound up with all the hopes, frustrations, and confusions of life."[66] Place is constructed partially through the built environment, partially through myth and narrative, and partially through human experience. Once a place, such as Delphi, has developed high status through cultural inscription on the land and in narrative, people will attempt to connect to that place in various ways. Here I address how Pindar develops Delphi as a place through etiological myth (*Paean* 8), how he manipulates the status of Delphi's temple of Apollo in the interest of his patrons (*Pythian* 7), and how his patrons themselves chose to commemorate their relationship to Delphi (*Pythian* 5). One geographical theme dominates this section, and that is the manner in which place, as Pred notes, is processual: "Place ... always involves an appropriation and transformation of space and nature that is inseparable from the reproduction and transformation of society in time and space. As such, place is not only what is fleetingly observed on the landscape, a locale, or setting for activity and social interaction. It also is what takes place ceaselessly, what contributes to history in a specific context through the creation and utilization of a physical setting."[67] Place, then, requires *praxis* as much as it requires *topos*.[68]

Pindar and his fellow Greek poets told numerous etiological myths that aided in constructing Delphi as a meaningful

65 Sheldrake 2001: 1.

66 Relph 1976: i. On "sense of place" see Relph 1997.

67 Pred 1984: 279. As Blanshard notes in this volume, "Forensic space ... relies less on specific architectural forms than on the demarcation of space through procedure" (Chapter 7: 251). For his discussion of the creation of forensic space through procedure, see Chapter 7: 251–55.

68 See Blanshard in this volume (Chapter 7: 279); also Dougherty in this volume (Chapter 4: 131–33), on the importance of literary poesis and performance for the construction of Athens.

place,[69] and Pindar creates an elaborate etiological myth for Delphi's temple of Apollo in the eighth *Paean*, which tells of several temples that preceded the seventh-century archaic stone temple.[70] Since the story of the four mythological temples is not preserved in any other known archaic sources, Pindar may have played a key role in the construction of the myth:[71]

> κλυτοὶ μάντι[ες] Ἀπόλλωνος,/ ἐγὼ μὲν ὑπὲρ χθονός/ [ὑ]π̣έρ τ' ὠκεα⌊νο⌋ῦ/Θε̣μιδ̣ό̣ς τ' ἐπὶ̣[(desunt vv. 5–50, restant frustula vv. 51–61) Ἰυ̣γ[γ . . .] / ναόν· τὸν μὲν Ὑπερβορ[έοις . . .]/ ἄνεμος ζαμενὴς ἔμ<ε>ιξ[. . .]/ ὦ Μοῖσαι· το<ῦ> δὲ παντέχ[νοις]/ Ἁφαίστου παλάμαις καὶ Ἀθά[νας]/ τίς ὁ ῥυθμὸς ἐφαίνετο;/ χάλκεοι μὲν τοῖχοι χάλκ[εαί] θ' ὑπὸ κίονες ἕστασαν,/ χρύσεαι δ' ἓ̈ξ ὑπὲρ αἰετοῦ/ ἄειδον Κηληδόνες./ ἀλλά μιν Κ̣ρόνο̣υ̣ π̣α̣ῖ̣ [δες]/κεραυνῷ̣ χθόν' ἀνοιξάμ[ε]ν̣ο̣[ι]/ ἔκρυπσαν τὸ [π]άντων ἔργων ἱερώτ[ατον]// γλυκείας ὀπὸς ἀγασ̣[θ]έντες,/ ὅτι ξένοι ἔφ[θ]<ι>νον/ ἄτερθεν τεκέων/ ἀλόχων τε μελ[ί]φρονι αὐδ[ᾷ̣ θυ]μὸν ἀνακρίμναντες· επε̣[. . .]/ λυσίμβροτον παρθενίᾳ̣ κε̣[. . .]/ ἀκηράτων δαίδαλμα [. . .]/ ἐ̣νέθηκε δὲ Παλλὰς ἀμ[. . .]/ φωνᾷ̣ τά τ' ἐόντα τε κα[ὶ] πρόσθεν γεγενημένα [. . .]ται Μναμοσύνᾳ̣[. . .]παντα σφιν ἔφρα[σ.ν. . . .

> Famous seers of Apollo, I (have come?) over the land, over the ocean, and to (the shrine?) of Themis . . . (gap of some 50 lines) Charmers temple. The one a furious wind

69 See Fontenrose 1974.

70 A fire destroyed the archaic temple in 548 BC. The "Alkmaionid" temple, built c. 510 BC, replaced the archaic temple, but, in 373 BC, an earthquake destroyed the Alkmaionid temple. The visitor to Delphi today sees the foundations not of the Alkmaionid temple but of its successor, which, for the most part, was completed by 305 BC. For discussion and reference to further bibliography, see Scott 2010: 56–59, 65–66, 114, 118.

71 Pausanias too provides an account of the temples (10.5.12). On the myth of the temples of Apollo see Sourvinou-Inwood 1979. The story of Apollo's temple is told also in the *Homeric Hymn to Apollo* (ll. 287–99), in which the temple of Apollo was constructed by Trophonios and Agamedes (Pindar mentions Trophonios and

> brought to the Hyperboreans ... O Muses. But of the other, what arrangement was shown by the all-fashioning skills of Hephaistos and Athena? The walls were of bronze and bronze columns stood in support, and above the pediment sang six golden Charmers. But the children of Kronos split open the earth with a thunderbolt and buried that most holy of all works, in astonishment at the sweet voice because strangers were perishing away from their children and wives as they suspended their hearts on the honey-minded song ... the man-releasing contrivance (?) of undamaged ... to the virgin ... and Pallas put (enchantment?) into their voice and Mnemosyne declared to them all the things that are and happened before ... (lines 1–4, 62–85)

Pindar asserts that multiple temples preceded the first stone temple at Delphi and he thereby builds a rich heritage for Delphi and Apollo's temple. Furthermore, as Sourvinou-Inwood notes, Pindar integrates the vegetal, the animal, the human, the heroic, and the divine spheres into the narrative of his various temples.[72] The integration of all known realms of life, then, constructs Delphi as a particularly powerful place. As Tuan notes,

> In studying how places at all scales have come into existence, geographers have focused almost exclusively on material processes and socioeconomic forces, without raising, explicitly, the role of language ... taking language seriously [however] has a number of intellectual consequences or rewards. It enables us to understand the *process* of place-making better by recognizing a force previously neglected, if not wholly ignored. It helps us to understand the *quality* (the personality or character) of place better, for that quality is imparted by, along with

Agamedes in a fragmentary Isthmian ode (fr. 2)). In contrast to the version found in the *Homeric Hymn to Apollo*, Pindar's eighth *Paean* develops a more elaborate narrative that tells of several temples that preceded the archaic stone temple.

72 Sourvinou-Inwood 1979.

> visual appearance and other factors, the metaphorical and symbolic powers of language.[73]

By connecting Delphi with time immemorial and by linking the full spectrum of life together at Delphi, Pindar, through language, creates a sense of place and a heritage for Delphi that few, if any, places could rival.

The first temple, having the shape of a hut, was made of laurel branches that came from the valley of Tempe, and the importance of laurel from Tempe in Apollo's cult at Delphi is well attested.[74] The papyrus on which the paean is preserved does not include narration of the first temple, but, as Rutherford notes, "We would expect the first temple to have been mentioned."[75] At Eretria, a temple made of laurel was dedicated to Apollo Daphnephoros, and a temple made of laurel at Delphi, then, need not have been pure myth; there is no archeological evidence, however, for a laurel temple at Delphi.[76] With laurel, Pindar integrates the vegetal realm into his ode, and Sourvinou-Inwood argues that "the laurel and the laurel temple take us ... away from the anthropomorphic sphere and from culture into wild nature. It is perhaps significant in this respect that nothing is said about who constructed this temple, or how it was done."[77] The primal character of vegetation, then, links the earliest periods of time with Apollo's first temple at Delphi.

Since laurel is a genuine building material, Pindar more clearly enters the realm of myth when bees construct Apollo's second temple out of birds' feathers and beeswax.[78] The section of the papyrus in which this temple was mentioned has been lost, but Pindar surely mentioned it, because, when the papyrus is legible, the narrator asserts that a strong wind sent

73 Tuan 1991: 692, 694.

74 For discussion and reference to further bibliography, see Sourvinou-Inwood 1979: 233–34.

75 Rutherford 2001: 218.

76 For discussion, see Sourvinou-Inwood, 1979: 234–38.

77 Sourvinou-Inwood 1979: 248.

78 Bees play an important role in Pythian myth; see Sourvinou-Inwood, 1979: 240–44.

this temple to the Hyperboreans. With the birds and the bees, Pindar moves the temple from the vegetal realm into the realm of animals, since wax derives from bees and feathers cover the bodies of birds.

Pindar integrates a cultural advance with the temple of bronze and, therewith, he introduces the gods into the production of place at Delphi, since Athena and Hephaistos built this temple. Athena and Hephaistos built the third temple with bronze walls and bronze pillars and added singing beings made of gold, known as Keledones ("Charmers"), as akroteria for the temple. The Keledones entranced men with their song, and, because the Keledones had a deleterious effect upon men, the gods hid this third temple under the earth.[79] This bronze temple, made out of a material that was widely used in Greek temple building, brings the temple of Apollo into a more historical context,[80] but Pindar squarely places this temple too in the realm of myth by having Athena and Hephaistos build the temple and by placing the Keledones on its roof. The integration of bronze adds cultural sophistication into the themes of the ode, due to the need to understand smelting to construct bronze.

The papyrus of *Paean* 8 breaks off before the mention of the fourth temple, but there is no doubt that Pindar went on to narrate it.[81] Trophonios and Agamedes built the fourth temple out of stone, and this fourth temple can be equated with Delphi's first historical temple.[82] With Trophonios and Agamedes and with a stone temple at Delphi, Pindar's myth begins to overlap with the myth of the construction of Apollo's temple as preserved in the *Homeric Hymn to Apollo* (ll.294–99).

79 As Pausanias notes, the Keledones are modeled off the Homeric Sirens. For discussion see Sourvinou-Inwood 1979: 244–46.

80 On bronze in temple architecture, see Sourvinou-Inwood 1979: 246.

81 See Rutherford 2001: 223.

82 For shorter lists (of the temples of Delphi) see Arist. fr. 3 Rose and Strabo 9.3.9. The myth may also have been mentioned in Callimachus' *Aetia* fr. 118 (cf. Pfeiffer ad loc.). The alternative versions have been gathered by Rutherford 2001: 231–32.

Trophonios and Agamedes are heroes, and in the Homeric Hymn, as well as in Pindar's paean, they fulfill an intermediary role between gods and men. In the Homeric Hymn, Apollo, Trophonios, Agamedes, and the "tribes of men" participate in building Apollo's temple, and the myth thereby integrates the divine, the heroic, and the human sphere.

Paean 8 is a good example of the importance of time, space, and narrative in place-making. As Rutherford remarks, "we see that Pindar aimed to include all spheres of nature in the construction of the temple."[83] Pindar intertwines stone, vegetation (laurel), natural materials (honey and feathers), and metals (gold and bronze), while linking gods, heroes, and humans in a temporal continuum that links the present with prehistory. As Pred notes, "As place-specific biographies are formed through social reproduction, and as place-specific social reproduction occurs through the formation of biographies, the physical environment is perpetually transformed. The transformation of the physical world is inseparable from the becoming of place."[84] Pindar forms Delphi's biography through several generations of temple building for Apollo, and, with every rebuilding of a temple, Delphi becomes more saturated with Apolline history. Delphi, then, becomes richer as a place with each of Pindar's temples.

While Pindar's eighth *paean* provides a mythological celebration of Apollo's temples at Delphi, *Pythian* 7 uses Apollo's temple as a symbol of socioeconomic power for Megakles of Athens, a member of the wealthy and politically influential Alkmaeonid clan. Megakles was ostracized from Athens in the early 480s and was most likely in exile at the time that Pindar composed *Pythian* 7.[85] After a fire destroyed the archaic

83 Rutherford 2001: 224.

84 Pred 1984: 287.

85 For discussion, see Athanassaki 2011: 238–43. This was not the first time that an Alkmaeonid was removed from Athens. Herodotus mentions four occasions of their exile in the sixth century. The Alkmaeonids, when exiled by the Peisistratids, may have gone to Phokis, though there is no direct evidence for this (Camp 1994: 7–12).

temple of Apollo, the Alkmaeonids took the contract to rebuild the temple; the contract called for the temple to be rebuilt in limestone, but the Alkmaeonids famously rebuilt the façade in Parian marble (Herodotus 5.62; cf. Arist. *Ath. Pol.* 22.5). According to Herodotus, the Alkmaeonids rebuilt the temple because they hoped to influence political affairs in Athens. The Alkmaeonid temple plays a key role in the brief ode, which may be cited in full:

Κάλλιστον αἱ μεγαλοπόλιες Ἀθᾶναι
προοίμιον Ἀλκμανιδᾶν εὐρυσθενεῖ
γενεᾷ κρηπῖδ' ἀοιδᾶν ἵπποισι βαλέσθαι.
ἐπεὶ τίνα πάτραν, τίνα οἶκον ναίων ὀνυμάξεαι
ἐπιφανέστερον
Ἑλλάδι πυθέσθαι;

πάσαισι γὰρ πολίεσι λόγος ὁμιλεῖ
Ἐρεχθέος ἀστῶν, Ἄπολλον, οἳ τεόν
δόμον Πυθῶνι δίᾳ θαητὸν ἔτευξαν.
ἄγοντι δέ με πέντε μὲν Ἰσθμοῖ νῖκαι, μία δ' ἐκπρεπής
Διὸς Ὀλυμπιάς,
δύο δ' ἀπὸ Κίρρας,
ὦ Μεγάκλεες,

ὑμαί τε καὶ προγόνων.
νέᾳ δ' εὐπραγίᾳ χαίρω τι· τὸ δ' ἄχνυμαι,
φθόνον ἀμειβόμενον τὰ καλὰ ἔργα. φαντί γε μάν
οὕτω κ' ἀνδρὶ παρμονίμαν
θάλλοισαν εὐδαιμονίαν τὰ καὶ τὰ φέρεσθαι.

> The great city of Athens is the fairest prelude to lay down as a foundation for songs to honor the mighty race of the Alkmaeonids for their horses. For what fatherland, what house can you inhabit and name with a more illustrious reputation in Hellas? None, for among all the cities travels the report about Erechtheus' citizens, Apollo, who made your temple in divine Pytho splendid to behold. Five victories at the Isthmos prompt me, as does one outstanding Olympic festival of Zeus and two victories at Kirrha. O Megakles, belonging to your family and

> forebears. I rejoice greatly at your recent success, but this grieves me, that envy requites your noble deeds. Yet they say that in this way happiness which abides and flourishes brings a man now this, now that.

Pindar opens the ode with architectural imagery, referring to Athens as a "*krepis* for songs" (κρηπῖδ᾽ ἀοιδᾶν). A *krepis* is an architectural foundation, and Pindar's metaphor for Athens serving as an ideal "foundation," that is, beginning of the ode, has received multiple interpretations. Neer suggests that the *krepis* resonates with the foundation course of stones for the treasury of the Athenians that would currently be visible at Delphi while it was under construction, while Steiner suggests that the opening "glances toward the numerous equestrian statues on the Athenian Acropolis dedicated by victors in racing events, monuments lavish in their scale and design, and visible testaments to the wealth of those families able to finance a chariot in the Games."[86] While Neer and Steiner both pull away from the temple of Apollo to explain the architectural metaphor, Athanassaki argues, "in so far as architectural metaphor points to equestrian sculptures, the primary reference is to the magnificent equestrian complex of the East pediment of the Alcmaeonid temple of Apollo."[87] Following Neer, I would suggest that the *krepis* stresses the importance of an architectural foundation, and, following Athanassaki, I would suggest that we should read the *krepis* in reference to the temple of Apollo. Nothing in Pindar's ode activates a connection with the Athenian acropolis or with the treasury of the Athenians, however, and because a *krepis* is a foundation, it does not trigger associations with the pediment of Apollo's temple.[88] With the architectural metaphor, then, I would suggest that Pindar uses an image that proleptically resonates with the importance

86 Neer 2004: 87; Steiner 1993: 166–67.

87 Athanassaki 2011: 246.

88 Nonetheless, Pindar's reference to the temple of Apollo later in this ode, following Athanassaki, does encourage the audience to imagine the temple of Apollo generally, and the audience will most obviously visualize the Alkmaeonid façade, including its pediment.

of monumental architecture, Apollo's temple, in the middle of the ode:

> πάσαισι γὰρ πολίεσι λόγος ὁμιλεῖ
> Ἐρεχθέος ἀστῶν, Ἄπολλον, οἳ τεόν
> 11/12 δόμον Πυθῶνι δίᾳ θαητὸν ἔτευξαν.

> for among all the cities travels the report about Erechtheus' citizens, Apollo, who made your temple in divine Pytho splendid to behold.

Although we may have expected Pindar to refer to his patron's family, the Alkmaeonids, as the builders of the temple, due to their important role in funding the temple, Pindar refers to the builders of Apollo's temple as the "citizens of Erechtheus" (Ἐρεχθέος ἀστῶν, l.10). As Athanassaki notes, "What is ... striking about this song is that the restoration of the temple of Apollo in Delphi, which together with the Alcmaeonid victories secures the pan-Hellenic visibility of Athens, is attributed to the citizens of Athens as a whole, at least at first sight."[89] By referring to the builders of the temple as the citizens of Erectheus, when the builders were the family of Megakles, Pindar makes a clear statement regarding Megakles' *polis* identity: Megakles is an Athenian and, whether he is still in exile or not at the time of the performance of this ode, his appropriate place is among his fellow citizens.

In addition to raising important questions regarding the power of place (as Pindar's use of the temple of Apollo in this ode bolsters the importance of the Alkmaeonids at Athens), then, this ode also plays on the issue of displacement and rootedness. Pindar's praise of Athens may at first seem jarring, since Megakles had been ostracized recently from Athens, but Pindar weaves the Alkmaeonids into the civic texture of Athens to such a degree that the Alkmaeonids are portrayed as a natural limb of the body of Athens. Accordingly, Megakles' displacement from his home *polis* may be read as an "unnatural" amputation of the civic body and, as the

[89] Athanassaki 2011: 236–37.

anthropologists Feld and Basso note, "displacement is no less the source of powerful attachments than are experiences of profound rootedness."[90] It is Megakles' absence, whether completed or ongoing, that makes his rightful presence in Athens all the more noticeable.

Pindar focuses in the beginning of the ode on the importance of Athens as an appropriate foundation for song and in the middle of the ode on the importance of Erechtheus' citizens, that is the Alkmaeonids, for the reconstruction of Apollo's temple. One cannot separate the city from its citizens, however, and the importance of the Alkmaeonids to Athens is clearly, though tactfully, referenced in the ode. As Athanassaki concludes, "the subordination of Alcmaeonid illustrious deeds to the glory of Athens should ... be seen as an attempt to win the good-will of the Athenians,"[91] and therewith Pindar positions Megakles among his fellow Athenians. Pindar exploits the temple of Apollo, then, to bolster the importance of his patron to Athens, and this ode is a good example of how a prestigious place, Delphi with its temple, can be manipulated in the interest of power.

Similarly, in *Pythian* 5, Pindar bolsters his patron's power and god-granted importance through Delphi. *Pythian* 5 begins with praise for Arkesilas IV, his wealth, and his noble pursuits, and Pindar, accordingly, chooses his themes well, while praising the tyrant in his home *polis*, Cyrene.[92] At lines 30–33, Pindar exhorts Arkesilas to hold in esteem his charioteer, Karrhotos, who enabled the victory for Arkesilas and who crowned Arkesilas thereafter:[93]

ἀλλ' ἀρισθάρματον
ὕδατι Κασταλίας ξενωθεὶς γέρας ἀμφέβαλε τεαῖσιν κόμαις,

[90] Feld and Basso 1996: 11.

[91] Athanassaki 2011: 243.

[92] The narrator envisions the performance of this poem in Cyrene (ll. 23–26).

[93] The reference to Karneian Apollo in line 79 (and further representation of cultic activity at Cyrene) has encouraged scholars to suggest

ἀκηράτοις ἀνίαις
ποδαρκέων δώδεκ' ἄν δρόμων τέμενος.

> But instead, after receiving hospitality by the water of Kastalia, he placed around your hair the prize for the first-place chariot, won with his reins intact in the sanctuary with its twelve swift-footed courses.

Pindar creates a tableau vivant for the audience so they can visualize Karrhotos beside Kastalia and can visualize Karrhotos crowning Arkesilas with the crown of laurel for his chariot victory. Thereafter, Pindar refers to the dedication of charioteering equipment at Delphi[94]:

κατέκλασε γὰρ ἐντέων σθένος οὐδέν· ἀλλὰ κρέμαται
ὁπόσα χεριαρᾶν
τεκτόνων δαίδαλ' ἄγων
Κρισαῖον λόφον
ἄμειψεν ἐν κοιλόπεδον νάπος
θεοῦ· τό σφ' ἔχει κυπαρίσσινον
μέλαθρον ἀμφ' ἀνδριάντι σχεδόν,
Κρῆτες ὃν τοξοφόροι τέγεϊ Παρνασσίῳ
καθέσσαντο μονόδροπον φυτόν.

> For [Karrhotos] broke none of his strong equipment, but it is hung in dedication – all that ornate handiwork of skilled craftsmen which he drove past the hill of Krisa on his way to the hollow valley of the god. And so, the

that this poem was performed at the Cyrenaian Karneia, a festival for Apollo, but it is equally possible that Pindar integrates the atmosphere of the Karneian festival into his ode for Arkesilas without any intention of it being performed at the Karneia. On the topography of Cyrene, related cultural-historical contextualization for *P.* 5, and the suggestion that the poem was composed for performance at the Karneia, see particularly Krummen 1990: 98–151.

94 It is unclear what the ἐντέων σθένος is. Pindar's language (35–36) suggests a substantial dedication. I presume that it is some equipment such as chariot wheels. If both chariot wheels were dedicated, this could explain the phrase "as much as he was driving." Nicholson (2005, 44, 228) suggests that it is the whole chariot that was dedicated, but, as he notes, this would be unparalleled. For dedications of chariot wheels at Delphi, see, for example, Taplin 2007, 61.

> shrine of cypress wood holds it beside the statue hewn from a single trunk, which the bow-bearing Cretans set up in the chamber of Parnassos.

The crowning of Arkesilas and the dedication of the charioteering equipment are relatively surprising in this ode, since Pindar generally shows little interest in specific details of a victory and its concomitant celebration.[95] In this ode, however, Pindar even apostrophizes Karrhotos (l.45) and congratulates him on not falling in the chariot race at which forty other drivers fell (ll. 49–51). It is unclear, however, whether 'fell' here refers to chariot crashes or serves as a metaphor for 'lost.'[96]

Although the description of the place where Karrhotos made his dedication is relatively detailed, it has not allowed scholars to establish unequivocally the place of the dedication at Delphi.[97] Pindar says that a shrine of Cypress wood holds the dedication, that there was a wooden statue housed inside the shrine, and that Cretans dedicated the statue within the shrine. Since the Cyrenaeans do not seem to have built a treasury at Delphi until the mid-fourth century BC, the dedication of Karrhotos should have been made elsewhere.[98] Scholars have suggested that the dedication was placed in a building constructed to house the offering, that it was placed in a "Kyrenaic shrine," or that Pindar's representation of dedication suggests that there was a treasury of the Cyrenaeans at Delphi already in the early fifth century.[99] Roux suggests that the reference

95 Bacchylides, on the other hand, frequently uses references to the contest and victory in manners that reenact the celebration and add vividness to his texts. See B. 5.37–41, 9.27–38, and particularly 10. 21–26. See also Carey 1980: 227–28.

96 Nicholson (2005, 43) takes Pindar's language literally.

97 The topic is unaddressed in Gildersleeve 1885; Krummen 1990; and Bernardini et al. 1995.

98 On the Cyrenaean treasury see Partida 2000: 147–72.

99 Building constructed to house the offering: Wilamowitz 1898: 381–82; Cyrenaic shrine: Fennell 1893: 200; early treasury of Cyrenaeans at Delphi: Puech 1962: 94.

is to a dedication made by Karrhotos within the temple of Apollo, positioned next to an ancient *xoanon* "wooden statue" of Apollo.[100] Although the place of dedication remains unclear, Pindar's intended audience would have known it. Accordingly, Pindar's depiction of the dedication in the ode provides Arkesilas with prestige not only due to Arkesilas' victory in the chariot race but also due to the votive that remains at Delphi as an eternal reminder of Arkesilas' accomplishment at Delphi.

As a speech act, Pindar's ode re-performs the dedication of the charioteering equipment and immortalizes Arkesilas' dedication at all re-performances of the ode. The equipment, furthermore, changes meaning as it moves from contest to celebration – what is first used to race horses is later used as a sign to construct social status. The equipment, accordingly, provides a provocative example of how objects can shift their meaning as they move through their life cycle.[101] Pindar's depiction of Arkesilas' dedication reaffirms Arkesilas' god-granted place of authority in the cosmos, and if the audience of Pindar's ode does not notice how Pindar uses Delphi, then they may be unaware that Pindar's presentation of Arkesilas' victory has reaffirmed Arkesilas' tyrannical authority at Cyrene.

CONCLUSION

In the first section, following Tuan and Felson, I emphasized how space is experienced through the body and how Pindar constructs space to allow vicarious transport to Delphi in *Pythian* 6. As Tuan remarks,

> Dynamism characterizes the feel of the visual-aural zone [i.e., of human sense perception], and this sense of a lively

[100] "Le poète nous apprend en effet que le char de Carrhôtos, aurige d' Arcèsilas, a été consacré dans le temple d' Apollon, suspendu à la charpente, non loin d' une antique statue de bois installée par les Crétois et qui, dans ce contexte, ne peut avoir été que la primitive statue de culte" Roux 1962: 368. See also Morris 1992: 44–45.

[101] See Gosden and Marshall 1999.

> world is the result of sound as much as spatial displacements that can be seen. When we turn from the distant visual space to the visual-aural zone, it is as though a silent movie comes into focus and is provided with sound tracks. Next to our body is the affective zone, which is accessible to the senses of smell and touch besides those of sight and hearing. In fact, the relative importance of sight diminishes in affective space: to appreciate the objects that give it its high emotional tone our eyes may even be closed.[102]

Pindar introduces the visual zone with images of procession and treasuries at Delphi. He introduces the aural zone with the imperative, "listen," with the adjective, "loud-rumbling," as well as with the stream of the voices of his chorus. He introduces the affective zone by having his audience feel the journey to the temple of Apollo and feel the rumbling earth. Importantly, Tuan emphasizes that once we are within the affective zone that the importance of sight diminishes. Similarly, the importance of actually being at Delphi and seeing first hand the phenomena that Pindar narrates diminishes because Pindar provides suitable vicarious transport for us to visualize and experience the absent. Pindar's ode, then, also allows for vicarious pilgrimage, because, with every performance, it brings Delphi to the place of performance.

In the second section, I focused on landscape, which, as Mitchell suggests, "is best understood as a medium of cultural expression."[103] Pindar's evocations and narrations that individually create senses of place and performative spaces for Delphi, when agglomerated, create a landscape idea, and it is through a landscape idea that representations of Delphi and the agonistic performances that occurred there become meaningful expressions of the dominant Greek culture that Pindar supports in his commissioned odes. As landscapes of privilege, Panhellenic landscapes, such as Delphi, implicitly reaffirm the social *status*

[102] Tuan 1975: 224.
[103] Mitchell (1994) 2002: 14.

quo and marginalize the unprivileged and excluded (e.g., slaves/women/non-Greeks),[104] and, as Rose remarks, "If Pindar often gives modern readers the impression that he speaks from and to a homogenous and untroubled aristocratic world, that is in part at least the mark of his triumph as a professional ideologue."[105] Pindar's Delphic landscape idea, moreover, reinforces spiritual bonds between Pindar's elite patrons and divinities (such as Kastalia) and affirms a god-granted good fortune for individual victors.[106] As Elliger notes, Pindar is primarily interested in Delphi as a landscape of power, not as a landscape of beauty, and, as we have seen, Pindar fits Kastalia squarely into this ideological matrix.[107]

Powerful places are powerfully useful, and Pindar did not miss the opportunity to use Delphi as a powerful place, as discussed in the third section. Pindar exploits the value inherent in the temple of Apollo through his construction of several mythical temples to precede the historic one in the eighth *Paean*, thereby providing Apollo's temple with deep heritage. This reminds us, following Pred, that place develops its personality through time.[108] Furthermore, as a reaffirmation of their status, the Alkmaeonids and Arkesilas IV inscribed their power at Delphi, through the Alkmaeonid temple of Apollo and

104 For a recent, thoughtful discussion of Panhellenism, see Scott 2010: 250–73.

105 Rose 1992: 144.

106 Cf. "Landscape is a medium not only for expressing value but also for expressing meaning, for communication between persons – most radically, for communication between the Human and the non-Human," Mitchell (1994) 2002: 15.

107 Elliger 1975: 208. It is helpful to think of Pindar's landscapes in relation to other Greek literary landscapes. Pindar does not show an interest in the potential beauty of landscape (Cf. Steiner 1986: 89); a romantic interest in the land had not yet developed in Greek literature (On Plato's role in the development of Greek landscape aesthetics; see Parry 1957: 15–20). Given the long tradition of beauty and aesthetics as topics of contemplation in both visual and literary depictions of the land cross-culturally, the absence of "the beautiful" in the construction of Delphic landscapes specifically, and epinician landscapes more broadly, is remarkable.

108 Pred 1984.

through Arkesilas' dedication of his equipment, and Pindar's narration of these on-site investments in Delphi adds further social capital for his patrons. Again following Pred, we see that place requires *praxis* as much as *topos* to give it meaning, and Pindar and his patrons invest in and perform at Delphi in a way that reaffirms their own importance and concurrently creates a sense of place for Delphi. Unsurprisingly, Pindar prefers the poetics of place to the science of place, and his representations are "accurate" not so much for the sake of descriptive realism as for the sake of connotation and symbolism.[109] Furthermore, as Thrift remarks, "like societies, places can be made durable but they cannot last."[110] Of course, Delphi no longer exists as it did for Pindar, for his patrons, or for the ancient Greeks broadly construed, but Pindar's Delphic sense of place can be recuperated to a certain degree from reading Pindar's odes today in relation to their original geographic context.

Bibliography

Acosta-Hughes, B. 2002. *Polyeideia: The Iambi of Callimachus and the Archaic Iambic Tradition*. Berkeley and Los Angeles.

Amandry, P. 1977. "Notes de topographie et d'architecture delphiques. VI. La fontaine Castalie." *Études delphiques (BCH Suppl. IV)*: 179–228.

1978. "Notes de topographie et d'architecture delphiques. VII. La fontaine Castalie (compléments)." *Bulletin de correspondance hellenique* 102: 221–41.

Athanassaki, L. 2011. "Song, politics and cultural memory: Pindar's *Pythian* 7 and the Alcmaeonid temple of Apollo." In *Archaic and Classical Choral Song: Performance, Politics and Dissemination*, ed. L. Athanassaki and E. Bowie. Berlin. 235–268.

Bernardini, P. A., E. Cingano, B. Gentili, and P. Gianninil. 1995. *Pindaro: Le Pitiche*. Rome.

Boeckh, A. (1821) 1963. *Pindari epiniciorum interpretatio latina cum commentario perpetuo*. Hildesheim.

109 Entrikin and Bourdoulay note in regard to Henri Lefebvre's historical geography *Pyrénées* that "when writing about place, Lefebvre appears to prefer connotation to denotation, the poetics of place over the science of place." Entrikin and Berdoulay 2005: 144.

110 Thrift 1999: 317.

Boehmer, F. 1891. *Pindars Sicilischen Oden nebst den Epizephyrischen*. Bonn.

Bommelaer, J.-F. 1991. *Guide de Delphes: Le Site*. Paris.

Bundy, E. 1962. *Studia Pindarica I. University of California Publications in Classical Philology*. Berkeley and Los Angeles.

Burton, R. W. B. 1962. *Pindar's Pythian Odes: Essays in Interpretation*. Oxford.

Bury, J. B. 1892. *The Isthmian Odes of Pindar*. London.

Calame, C. 2003. *Myth and History in Ancient Greece: The Symbolic Creation of a Colony*. Princeton.

Camp, J. McK. II. 1994. "Before Democracy: The Alkmaionidai and Peisistratidai." In *The Archaeology of Athens and Attica under the Democracy*, ed. W. D. E. Coulson, O. Palagia, T. L. Shear, Jr., H. A. Shapiro, and F. J. Frost. Oxford. 7–12.

Carey, C. 1980. "Bacchylides Experiments: Ode 11." *Mnemosyne* 33: 225–43.

2007. "Pindar, Place, and Performance." In *Pindar's Poetry, Patrons, and Festivals: From Archaic Greece to the Roman Empire*, ed. S. Hornblower and C. Morgan. Oxford. 199–210.

Clay, J. S. 1999. "Pindar's Sympotic Epinicia." *QUCC* 62: 25–34.

Cosgrove, D. 1998. *Social Formation and Symbolic Landscape*. Madison.

Dillon, M. 1997. *Pilgrims and Pilgrimage in Ancient Greece*. London.

Dougherty, C. 1993. *The Poetics of Colonization: From City to Text in Archaic Greece*. Oxford.

Duncan, J. and N. Duncan. 1988. "(Re)reading the Landscape." *Environment and Planning D: Society and Space* 6: 117–26.

Eckerman, C. 2011a. "Pindar's Pythian 6: On the Place of Performance and an Interpretive Crux." *RM* 153: 1–8.

2011b. "Pindar's Olympian 1.17 and Solo vs. Choral Epinician Performance." *Mnemosyne* 64: 83–85.

2012. "Was Epinician Poetry Performed at Greek Sanctuaries?" *GRBS* 52: 338–60.

Elliger, W. 1975. *Die Darstellung der Landschaft in der griechischen Dichtung*. Berlin.

Elsner, J. and I. Rutherford (eds.). 2005. *Pilgrimage in Graeco-Roman & Early Christian Antiquity: Seeing the Gods*. Oxford.

Entrikin, N. and V. Berdoulay. 2005. "The Pyrenees as Place: Lefebvre as Guide." *Progress in Human Geography* 29: 129–47.

Farnell, L. R. 1932. *The Works of Pindar*. London.

Feld, S. and K. Basso. 1996. "Introduction." In *Senses of Place*, ed. S. Feld and K. Basso. Santa Fe, NM. 3–11.

Felson, N. 1999. "Vicarious Transport: Fictive Deixis in Pindar's Pythian Four." *HSCP* 99: 1–31.

Fennell, C. A. M. 1893. *Pindar: The Olympian and Pythian Odes*. Cambridge.

Fontenrose, J. E. 1974. *Python: A Study of Delphic Myth and its Origins*. New York.

Gelzer, T. 1985. "*Mousa Authigenes*: Bemerkungen zu einem Typ Pindarischer und Bacchylideischer Epinikien." *MH* 42: 95–120.

Gildersleeve, B. L. 1885. *Pindar: The Olympian and Pythian Odes*. New York.

Gosden, C. and M. Yvonne. 1999. "The Cultural Biography of Objects." *World Archaeology* 31: 169–78.

Hardie, A. 1996. "Pindar, Castalia and the Muses of Delphi (The Sixth Paean)." *Papers of the Leeds International Latin Seminar* 9: 219–57.

Harrell, S. 2002. "King or Private Citizen: Fifth-Century Sicilian Tyrants at Olympia and Delphi." *Mnemosyne* 55: 439–64.

Heath, M. 1988. "Receiving the *komos*: The Context and Performance of Epinician." *AJP* 109: 180–95.

Hetherington, K. 2003. "Spatial Textures: Place, Touch, and Praesentia." *Environment and Planning A* 35.11: 1933–44.

Jones, P. 2005. *Reading Rivers in Roman Literature and Culture*. Lanham, MD.

Käppel, L. 1992. *Paian: Studien zur Geschichte einer Gattung*. Berlin.

Kong, L. 2001. "Mapping 'New' Geographies of Religion: Politics and Poetics in Modernity." *Progress in Human Geography* 25: 211–33.

Krummen, E. 1990. *Pyrsos Hymnon. Festliche Gegenwart und mythisch-rituelle Tradition als Voraussetzung einer Pindarinterpretation (Isthmie 4, Pythie 5, Olympie 1 und 3)*. Berlin.

Kurke, L. 1991. *The Traffic in Praise: Pindar and the Poetics of Social Economy*. Ithaca.

Larson, J. 2001. *Greek Nymphs: Myth, Cult, Lore*. Oxford.

Malkin, I. 2003. "Networks and the Emergence of Greek Identity." *Mediterranean Historical Review* 18: 56–74.

2011. *A Small Greek World: Networks in the Ancient Mediterranean*. Oxford.

Mitchell, W. J. T. (1994) 2002. *Landscape and Power*. 2nd ed. Chicago.

Morgan, C. 1990. *Athletes and Oracles: The Transformation of Olympia and Delphi in the Eighth Century BC*. Cambridge.

Morris, S. P. 1992. *Daidalos and the Origins of Greek Art*. Princeton.

Morrison, A. D. 2007a. *The Narrator in Archaic and Hellenistic Greek Poetry*. Cambridge.

2007b. *Performances and Audiences in Pindar's Sicilian Victory Odes*. London.

Neer, R. T. 2001. "Framing the Gift: The Politics of the Siphnian Treasury at Delphi." *ClasAnt* 20: 273–336.

2004. "The Athenian Treasury at Delphi and the Material of Politics." *ClasAnt* 23.1: 63–93.
2007. "Delphi, Olympia, and the Art of Politics." In *The Cambridge Companion to Archaic Greece*, ed. H. A. Shapiro. Cambridge. 245–64.
Nicholson, N. 2005. *Aristocracy and Athletics in Archaic and Classical Greece*. Cambridge.
Parke, H. W. 1978. "Castalia." *Bulletin de correspondance hellénique* 102: 199–219.
Parry, A. 1957. "Landscape in Greek Poetry." *YCS* 15: 3–29.
Partida, E. C. 2000. *The Treasuries at Delphi: An Architectural Study*. Jonsered, Sweden.
Pred, A. 1984. "Place as Historically Contingent Process: Structuration and the Time-Geography of Becoming Places." *Annals of the Association of American Geographers* 74: 279–297.
Puech, A. 1962. *Pindare, v. II, Pythiques*. Paris.
Purves, A. 2010. *Space and Time in Ancient Greek Narrative*. Cambridge.
Race, W. (ed.) 1997. *Pindar: Olympian Odes, Pythian Odes*. Cambridge, MA.
Radt, S. L. 1958. *Pindars zweiter und sechster Paian*. Amsterdam.
Relph, E. 1976. *Place and Placelessness*. London.
1997. "Sense of Place." In *Ten Geographic Ideas That Changed the World*, ed. S. Hanson. New Brunswick, NJ. 205–26.
Rodaway, P. 1994. *Sensuous Geographies: Body, Sense, and Place*. London.
Rose, P. 1992. *Sons of the Gods, Children of Earth: Ideology and Literary Form in Ancient Greece*. Ithaca.
Roux, G. 1962. "Pindare, le prétendu trésor des Crétois et l'ancienne statue d'Apollon à Delphes." *REG* 75: 366–80.
Rutherford, I. 2001. *Pindar's Paeans: A Reading of the Fragments with a Survey of the Genre*. Oxford.
2003. "The Prosodion: Approaches to Lyric Genre." In *Studi di filologia e tradizione greca in memoria de A. Colonna*, ed. F. Benedetti and S. Grandolini. Naples. 713–26.
Sack, R. A. 2003. *A Geographical Guide to the Real and the Good*. New York.
Scott, M. 2010. *Delphi and Olympia: The Spatial Politics of Panhellenism in the Archaic and Classical Periods*. Cambridge.
Sheldrake, P. 2001. *Spaces for the Sacred: Place, Memory and Identity*. Baltimore.
Slater, W. J. 1969. "Futures in Pindar." *CQ* 19: 86–94.
Sourvinou-Inwood, C. 1979. "The Myth of the First Temples at Delphi." *CQ* 29: 231–51.
Steiner, D. 1986. *The Crown of Song: Metaphor in Pindar*. New York.

1993. "Pindar's "Ogetti Parlanti." *HSCP* 95: 159–180.

Taplin, O. 2007. *Pots and Plays: Interactions Between Tragedy and Greek Vase-Painting of the Fourth Century B.C.* Los Angeles.

Thrift, N. 1999. "Steps to an Ecology of Place." In *Human Geography Today*, ed. D. Massey, J. Allen, and P. Sarre. Cambridge. 295–322.

Tuan, Y. -F. 1975. "Space and Place: Humanistic Perspective." *Progress in Geography* 6: 211–52.

1991. "Language and the Making of Place: A Narrative-Descriptive Approach." *Annals of the Association of American Geographers* 81: 684–696.

Wilamowitz-Moellendorff, U. von 1898. *Bakchylides*. Berlin.

1922. *Pindaros*. Berlin.

Wylie, J. 2007. *Landscape*. London.

2

SPACE AND LANDSCAPE IN XENOPHON'S *ANABASIS*

Tim Rood

"Look," said Hugh, "it must be Thesiger."

Coming towards us out of the great gorge where the river thundered was a small caravan like our own. He named an English explorer, a remarkable throwback to the Victorian era, a fluent speaker of Arabic, a very brave man ... as hard as nails ...

The ground was like iron with sharp rocks sticking up out of it. We started to blow up our air-beds. "God, you must be a couple of pansies," said Thesiger.

Eric Newby, *A Short Walk in the Hindu Kush*

They also held an athletic competition on the mountain where they were encamped, and they chose Dracontius the Spartiate (who had been banished from his home while still a boy for having accidentally stabbed another boy with his dagger and killed him) to be responsible for the racetrack and to supervise the contest ... The generals told him to show them where he had laid out the track. He pointed to the place where they happened to be standing and said: "This hill is perfect for running in any direction." "But how," they asked, "will they be able to wrestle on hard, shrub-covered ground like this?" And Dracontius replied: "It'll be a bit more painful for the one who is thrown."

Xenophon, *Anabasis*[1]

1 Quotes: Newby (1958) 2008: 248; Xen. *Anab.* 4.8.26. All references will be to the *Anabasis* unless otherwise indicated; translations are from Waterfield 2005, at times adapted to bring out particular linguistic points.

Xenophon's *Anabasis* offers rich pickings for the study of Greek representations of space and landscape. It is a detailed eyewitness account of the varied movements of an army of Greek mercenaries over the course of two years: a long march upcountry from Sardis to Mesopotamia; a retreat north through Kurdistan and Armenia to the Black Sea coast; a journey (partly by land, partly by sea) along the coast to Byzantium; a winter's campaigning in Thrace; and finally, a trip back across the Bosporus to Asia and then along the coast to Pergamum. Some readers have nonetheless been disappointed by the quality of the spatial information Xenophon supplies. Parts of the account have been hailed as picturesque,[2] but other sections seem formulaic. The way they present spatial information almost suggests secondhand use of a written source rather than autopsy. Xenophon rarely offers the sort of in-depth detail about landscape use that modern historians of the Achaemenid empire would like. This discussion, by focusing on the relation between space and landscape[3] in two sections of the *Anabasis* (the march to the sea and the march along the Black Sea coast), aims to provide some positive explanations for the varying modes of landscape description that Xenophon does employ: it will suggest that presentation of both space and landscape serves as an important part both of Xenophon's construction of an idealized image of his own generalship and of his exploration of (shifting patterns of) Greek cultural identity.

Some scholars have expressed doubt that Greek landscape descriptions offer insight into cultural identity. In a recent

2 See Rood 2010a: 82 and 2012.

3 There is, of course, a very large bibliography about these concepts as well as the related term "place" (e.g., Massey 1993; May 1996; Agnew 2005). "Space" is generally seen as a more abstract category than "place" (space as experienced) or landscape (space as viewed); much recent discussion focuses specifically on contemporary notions of place in an era of globalization. I use them as heuristic terms that can helpfully clarify spatial relations in ancient texts. I have treated one aspect (the marching formulae, especially the parasang) in Rood 2010b; I hope to return elsewhere to the different techniques (particularly for the presentation of rivers and cities) used for the march up-country.

wide-ranging article on "Landscape, Identity, and War," Svend Erik Larsen has argued that Greek depictions of war generally rely on "a conception of cultural identity without landscape" and that Xenophon's *Anabasis* in particular fits this pattern: "[L]andscapes are seen only as a means for the survival of the Hellenic army – a stock of provisions or as an impediment for their progress on their return ... a device to underline their thoughtful character and indefatigable prowess."[4] While there is much truth in this assessment, the Dracontius anecdote (cited in the second chapter epigraph) points, I suggest, to a deeper interrelation between the Greeks' view of landscape and their sense of their cultural identity – no less than Eric Newby's portrayal of Thesiger speaks to a deep interconnection between landscape and English identity. Significantly placed as the Greeks celebrate athletic games on their return to the Black Sea – that is, to the fringes of the Greek world – Dracontius' laconic words offer some sort of retrospective insight into the army's perceptions of the landscapes they have encountered during their retreat from Mesopotamia.

I return to Dracontius (and Thesiger) in my conclusion. The main body of this chapter consists of two distinct but interrelated sections, bound together by a narratological concern with voice and focalization and by a focus on cultural identity. Both examine the interplay between the spatial information and landscape descriptions offered by Xenophon the narrator and the speeches and perspectives of Xenophon the character (from now on "Xenophon") and of other characters. Xenophon's presentation of himself in the third person (possibly accompanied by the use of a pseudonym) has often been seen as a vehicle for tactful self-praise.[5] There has, however, been relatively little detailed analysis of the interaction between the viewpoints of the narrator and the character "Xenophon." My discussion of

4 Larsen 2004: 473, 474–75. For the importance of strategic and tactical space, compare Rambaud 1974 and Riggsby 2006: 24–28 on Caesar.

5 See Pelling 2013 for discussion of existing scholarship on this issue (with an important reminder of the need to historicize the concept of pseudonymity) and an excellent extensive comparison with Caesar.

Xenophon's presentation of space and landscape illustrates the complexity that arises from this dialectic. First, I show how the perspectives of narrator and characters interact in constructing an image of the Greeks' conquest of spatial difficulties and how this image can be read as part of an intellectual and emotional engagement with an archetypal Greek narrative of space. Next, by analyzing the differences between routine march sections in which Xenophon generally pays little attention to terrain and military passages in which the army's progress is disrupted and Xenophon's topographical focus is much sharper, I suggest that Xenophon boosts his own authority as a model general by the way he controls access to geographical description. I go on to relate Xenophon's presentation of space more strongly to his implied depiction of the mercenaries as emblematic of the strengths and weaknesses of Greek values. We may start, however, by exploring the ways in which issues of cultural identity are staged as Xenophon presents the Greeks stranded in Asia after Cyrus' death.

Narrator and Character Space: From *Aporia* to *Euporia*

Central to Xenophon's plotting of the Greeks' encounter with Persian space is an emotional passage in which he brings out how desperate the Greeks' position is after Cyrus' death in battle and the subsequent arrest of five of their generals:

> The Greeks were in great difficulty as they reflected that they were at the doors of the king; all around them there were countless hostile tribes and cities; there was no longer anyone who would sell them provisions; they were at least 10,000 stades from Greece; there was no guide to show them the way; there were uncrossable rivers blocking their route home. (3.1.2)

Xenophon here focalizes the Greeks' difficulty through the soldiers themselves. The troops are oppressed by the human and geographical obstacles they face. Their position is defined by negation: they were "in great difficulty" (ἐν πολλῇ δὴ ἀπορίᾳ), that is, there is no way out (the literal sense of ἀπορία). The stress on negation continues when Xenophon describes the troops as weighed down by "depressing thoughts" (ἀθύμως ἔχοντες): the word ἀθύμως has the same alpha-prefix as ἀπορία, again signifying privation. The description of the Greeks as surrounded on all sides (κύκλῳ ... πάντῃ) and facing a hostile landscape of "uncrossable rivers" (ποταμοὶ ... ἀδιάβατοι) also marks a return to language applied to their position earlier in their march: in the battle of Cunaxa, the Greek general Clearchus refuses to move the Greeks opposite the Persian center for fear of being "encircled from both sides" (1.8.13: κυκλωθείη ἑκατέρωθεν); immediately after the battle, Phalinus, a Greek in Persian service, tells the mercenaries that the king "has you in the middle of his territory, caught between uncrossable rivers" (2.1.11:ἔχων ἐν μέσῃ τῇ ἑαυτοῦ χώρᾳ καὶ ποταμῶν ἐντὸς ἀδιαβάτων[6]).

Xenophon's description of the Greeks' spatial difficulties draws on intertextual as well as intratextual echoes. Surrounded by impassable rivers, the troops have figuratively followed the steps of the Homeric Odysseus down to the Underworld. When Odysseus meets his mother there, she expresses astonishment at how he has made the journey alive: "My child, how did you come here beneath the fog and the darkness / and still alive? All this is hard for the living to look on, / for in between lie the great rivers (μέσσῳ γὰρ μεγάλοι ποταμοί) and terrible waters / that flow, Ocean first of all, which there is no means of crossing / on foot, not unless one has a well-made ship" (*Od.* 11.155–59, trans. Lattimore). Odysseus' mother is astonished at how he has managed to cross the great rivers between the Underworld and the land of the living – rivers that cannot be crossed on foot. So too Xenophon's Greeks are faced with the problem of

6 The adjective ἀδιάβατος occurs in these two passages in the *Anabasis*; it is first attested in extant literature in Xenophon, who also uses it at *Hell.* 5.4.44 (uncrossable ravine), again in connection with ἀπορία. For rivers as obstacles within the *Anabasis*, cf. Baslez 1995: 84–85; Roy 2007: 67.

"uncrossable rivers blocking their route home" (ποταμοὶ δὲ διεῖργον ἀδιάβατοι ἐν μέσῳ τῆς οἴκαδε ὁδοῦ). Unlike Odysseus, however, they have no ships to help them cross those rivers. Nor do they have any outside help to get them out. No surprise, then, that they no longer expect to see their "homes, parents, wives, and children" again (3.1.3).[7]

Another possible intertextual resonance is the escape drama, represented in surviving plays by Euripides, *Helen* and *Iphigenia in Tauris*. In both plays, female Greek characters trapped in distant lands (Egypt and the shores of the Black Sea, respectively) achieve escape by sea thanks to the fortunate arrival of male relatives.[8] The contrast with the position of Xenophon's Greeks is suggested by the fact that escape by land is expressly discussed in both plays, only to be rejected. Immediately after the recognition of Orestes, Iphigenia ponders how to find a path (πόρον, *IT* 876): "Should it be on land, not by ship but with flurry of feet? Then you'll come close to death as you go through barbarous tribes and impassable paths (ὁδοὺς ἀνόδους)." (*IT* 884–88, trans. Cropp). Similarly, Helen asks how they are to flee given that they "know nothing of the plains of this savage land (πεδίων ἄπειροι βαρβάρου γ' ὄντες χθονός)" (*Helen* 1042, trans. Morwood). Just as in the Euripidean plays, so too in Xenophon an alien and distant landscape of adventure serves as an affirmation of Hellenic resourcefulness. Xenophon's soldiers have to rely on themselves to escape, and they have large tracks of land to cross.

Xenophon's account of how there were "countless hostile tribes and cities" all "around" the army (κύκλῳ ... πάντῃ πολλὰ καὶ ἔθνη καὶ πόλεις πολέμιαι) can also be read against idealized portrayals of Greek military endeavors. A paradigmatic

7 By contrast with Odysseus, who does see his mother, if only in the Underworld, and does achieve the return home that the Greeks at this point despair of. Cf. also the description of Penelope at *Od.* 4.787–841: first sleepless and not touching food out of anxiety for Telemachus, then (like Xenophon – see discussion later in the book) visited by a dream when she does sleep. Another epic link is the danger of isolation on the battlefield (cf. e.g. *Il.* 11.470, 22.456).

8 Wright 2005: 204.

example appears in Herodotus' account of the prelude to the battle of Plataea, when the Phocians – characterized as unwilling medizers – have joined up with the Persians. The Persian general Mardonius first orders the Phocians to position themselves on the plain. Then the entire Persian cavalry arrives on the scene and makes as if to charge. At this point the Phocian commander appeals strongly to Greek superiority: "[L]et them learn that they are mere barbarians who have contrived to murder Hellenes who are *men*" (9.17.4). The Persian cavalry that surrounded (ἐκυκλώσαντο) them charges but suddenly turns away when the Phocians close ranks on all sides (9.18.1); Mardonius subsequently praises them as "brave men" (ἄνδρες ... ἀγαθοί, 9.18.3) – a phrase strongly resonant of Greek celebrations of their performances in the Persian Wars. This Phocian example is especially striking because a display of Greek valor is extracted from troops who are actually fighting on the Persian side. Similar descriptions of Greeks as encircled by non-Greeks are found in Herodotus' account of the key Persian Wars battles on both land and sea.[9] This language of encirclement follows easily from the Greek presentation of their numerical disadvantage: the Persian Wars were regularly presented as a combat between Greek virtue and Persian numbers.

Xenophon's language of encirclement draws not just on specific portrayals of the vast numbers of the Persians' army of invasion but also on a broader Greek geopolitical perception of the barbarians as a numerically superior encircling mass.

9 At Thermopylae, the Persians "come round to surround them on all sides" (7.225.3: περιελθόντες πάντοθεν περισταδόν); at Artemisium, in the first battle the Persians "surrounded the Greeks in a circle of their ships" (8.10.1: ἐκυκλοῦντο ... ἐς μέσον) and the Ionians perceive of the Greeks as "encircled" (8.10.2: περιεχομένους), while in the second the Persians "began to encircle" the Greeks (8.16.1: ἐκυκλοῦντο); at Salamis, verbs of encirclement are again used in both the middle, of the Persians encircling the Greeks (κυκλούμενοι 8.76.1, περιεκυκλοῦντο 8.78.1) and the passive, of the Greeks being encircled ("Aristides" at 8.79.4: περιεχόμεθα ... κύκλῳ; Themistocles at 8.80.2: περιεχόμεθα πανταχόθεν – the point in these speeches being to underline the necessity of fighting when the Peloponnesian contingents would have preferred to leave Salamis).

Herodotus makes a large population one of the hallmarks of barbarian peoples dwelling on the margins of the world.[10] His image was schematized by Ephorus, who portrays non-Greek peoples (Scythians, Celts, Ethiopians, Indians) implicitly surrounding the (less numerous) Greeks in the center: "[I]f the regions of the heavens and of the earth are divided into four parts, the Indians occupy that part from which Apeliotes blows, the Ethiopians that part from which Notus blows, the Celts the part on the west, and the Scythians the part from which the north wind blows."[11] The threat of barbarian neighbors was seen by Xenophon himself as a general Greek predicament: "[O]n the borders of most states dwell barbarians who trouble them" (*Poroi* 1.8).[12] Allowance must admittedly be made for the rhetorical context of this passage: Xenophon

10 Herodotean examples include 3.94.2 (Indians), 4.81.1 (Scythians), and 5.3.1 (Thracians). Cf. Fehling 1989: 187.

11 Ephorus, *FGrHist* 70 F 30 (from Strabo 1.2.28). Note also Andron of Halicarnassus frag. 7 Fowler, where the account of four daughters of Oceanus after whom the continents were named – Europe and Thraike by one mother, Asia and Libye by another – seems to suggest a divide of the inhabited world into northwest, northeast, southeast, and southwest, with Greece implicitly in the middle. Greece's centrality is explicit in Aristotle's *Politics*, where the Greeks are geographically in the middle of the colder, more spirited peoples of Europe and the (by implication warmer) peoples of Asia (see especially 1327b29–30: μεσεύει κατὰ τοὺς τόπους). Cf. Xen. *Poroi* 1.6: "One might reasonably suppose that the city [Athens] lies at the center (ἀμφὶ τὰ μέσα) of Greece, nay of the whole inhabited world. For the further we go from her, the more intense is the heat or cold we meet with"; the Hippocratic *Airs, Waters, Places*, by contrast, identifies the most fertile part of Asia as that "situated midway (ἐν μέσῳ) between the heat and the cold" (12.4), while also linking the fertility of Asia as a whole with its position "midway (ἐν μέσῳ) between the [winter and summer] risings of the sun" (12.3).

12 Cf. Polybius on the position of Byzantium: "As Thrace encompasses their territory (κύκλῳ περιεχούσης) so effectually as to extend from one sea to the other, they are engaged in perpetual and most difficult warfare with its inhabitants," which cannot be resolved by a single victory "owing to the great number (πλῆθος) of the chieftains and their followers" (4.45.1–2); or again the Polybian speaker who describes Macedonia as a "fence" protecting the Greeks (9.35.3–4: πρόφραγμα, προκαθημένων). Cf. also Dion. Hal. *AR* 4.25.3.

is here magnifying the advantages of Athens' supposed centrality between regions of hot and cold (see n. 11) by claiming that "the neighboring states of Athens are themselves remote from the barbarians." But the terms of Xenophon's praise of Athens merely underline the danger confronting the Greek army in Mesopotamia. Like Athens, they are figuratively in a center of sorts: Xenophon has earlier portrayed Cyrus presenting Persian power as extending to realms of northern cold and southern heat (1.7.6), and he repeats those terms in the *Cyropaedia* (8.6.21), in which he also defines the empire's limits to the west and to the east (limits imposed by water in one direction and by lack of water in the other) and adds that the elder Cyrus lived "in the middle" (8.6.22: ἐν μέσῳ) of this vast empire.[13] The Greeks, then, are also in a center between extremes of hot and cold, but they have no buffer to protect them from the imminent barbarian threat.

Xenophon's presentation of the external spaces of the Persian Empire surrounding the Greek soldiers makes them, in a sense, emblems of Greek identity *tout court*. This emblematic status is reinforced by a contrast with the way he describes the position of the Persian king. By speaking of the Greeks' perception that they were "at the doors of the king" (3.1.2: ἐπὶ ταῖς βασιλέως θύραις), he presents the king as contained within an internal space – unlike Greeks, who stereotypically undertake their public roles in the open air.[14] The phrase also, however, points to the Greeks' own dangerous proximity to the king.

13 The complexity of notions of geographical centrality is further underlined by the rivers "in the middle" (ἐν μέσῳ) of the Greeks' journey home at 3.1.2.

14 For other non-Greek internal spaces, cf. the tents of Cyrus (1.6.4, 1.7.8) and Tiribazus (4.4.21) and village houses temporarily occupied by the Greek soldiers (4.5.25–29, 7.4.14–16). No scene takes place within an ordinary Greek tent (at 1.2.17–18 and 1.5.12, characters return to their tents; at 3.3.1 the soldiers abandon their tents following the advice of "Xenophon" (3.2.2); cf. also 5.5.20, 6.4.19; Lee 2007: 122). Cf. also Callicratidas' complaints about being made to wait at Cyrus' doors at *Hell*. 1.6.7, 1.6.10. On Greek and "barbarian" conceptions of inside versus outside, see Purves' discussion in this volume (Chapter 3).

The danger is even greater because of two intratextual echoes. Anonymous soldiers had earlier complained that the Persian king was using negotiations as a mask for aggressive preparations: "[H]e's not willingly going to let us return to Greece with the news that, for all our small numbers, we defeated the king at his doors (ἐπὶ ταῖς θύραις αὐτοῦ), made him a laughing stock, and then left" (2.4.4). Here the phrase is part of a boast about the Greeks' achievement at Cunaxa, but it also carries the same sense of danger present in the Greeks' thoughts at 3.1. And the other recent use of the phrase suggests a much more direct danger: when the generals and captains visiting Tissaphernes had arrived "at his doors" (ἐπὶ θύραις ταῖς Τισσαφέρνους), the generals had been invited in and then arrested while the captains waiting "at his doors" (ἐπὶ ταῖς θύραις) had been killed on the spot (2.5.31).

The main function of Xenophon's sketch of the Greeks' position is to prepare for the introduction of their savior – the character "Xenophon" himself ("There was in the army a certain Xenophon, an Athenian ..."). Xenophon picks up his account of the Greeks' perception of their position (particularly the ideas of ἀπορία and encirclement) as he describes how "Xenophon" has a dream of a thunderbolt striking his father's house. On the one hand, he was minded to give the dream a positive interpretation, as "a great light from Zeus." But he was also afraid: "[T]he fact that the fire had cast its light all around (κύκλῳ) suggested that he might not be able to escape from the king's territory, but might be hemmed in on all sides by various difficulties (πάντοθεν ὑπό τινων ἀποριῶν)" (3.1.12).[15] This interpretation of the dream reinforces the account of the Greeks' spatial obstacles – thus prefiguring "Xenophon"'s own role as the Greeks' savior.

The light of the thunderbolt is thematically linked with the earlier narrative's use of darkness as a spatial metaphor. This image appears in a speech made by Clearchus when he meets Tissaphernes in an attempt to remove the air of mutual

[15] Cf. also the repetition of ἀπορία at 3.1.11.

suspicion that has hindered cooperation between the Greek mercenaries and the Persians since Cyrus' death: "With you, we easily travel every road, cross every river, and keep ourselves supplied with provisions. Without you, every road is shrouded in the darkness of our ignorance; every river is hard to cross ... and there is nothing more frightening than the uninhabited places" (2.5.9). As often, darkness is associated with a lack of knowledge (the words for sight and knowledge are associated). Clearchus' keenness to remove his fears paradoxically prepares the way for them to be fulfilled, and the metaphor of darkness is immediately realized in Xenophon's description of the anguished night after the seizure of the generals.[16]

Xenophon's stress on the Greeks' ἀπορία during that night also picks up the earlier narrative. The term ἀπορία appears when a dispute arises at Tarsus from the mercenaries' growing suspicion that they are being led against the Persian king instead of the Pisidians. Some point out "the difficulty (ἀπορία) of either staying or leaving without Cyrus' consent" (1.3.13); one man makes the (deliberately absurd) proposal that they should ask Cyrus to provide ships or a guide by land, or else seize the heights themselves (1.3.14), enabling the next speaker to point out that a guide given by Cyrus could lead them "into a position from which it will be impossible to escape" (1.3.17). There are also two uses of ἀπορία (as well as one of its opposite, εὔπορος) in Clearchus' speech to Tissaphernes (2.5.9). The term has a strong spatial dimension: it points to the Greeks' need to find a way (πόρος) back to their homeland. This spatial aspect is further picked up in a regular formula Xenophon applies to the Greeks' march: whereas on the march up-country Cyrus "progresses" (ἐξελαύνει) so many

16 For light/dark, cf. Iphigenia at Eur. *IT* 746, "I shall save you out of the dark rocks" – "a symbolic way of promising to see him safely out of the Black Sea to the Greek world" (Hall 1987: 429); Dodds 1960 ad loc. on Eur. *Bacch.* 608 (the "light" of deliverance); Ap. Rhod. *Argon.* 2.333, also (a literal instance) 4.296–97 ("a furrow of radiant light in the heavens marked the path they must take").

stages and parasangs, during the retreat the Greek "made their way" (ἐπορεύθησαν).[17]

The Greeks' difficulties after Cyrus' death are highlighted by contrast with the Persians' lack of spatial ἀπορία. Xenophon makes Tissaphernes play repeatedly on the πόρος-root as he explains to Clearchus why the Persians could easily destroy the Greeks:

> Do you perhaps think that we are short (ἀπορεῖν) of suitable places from which to launch an attack against you? Can't you see how vast these plains are and how arduous your journey across them is (διαπορεύεσθε) even under favorable conditions? Can't you see the huge mountain ranges you're going to have to cross (πορευτέα), where we could occupy the passes before you got there and stop you getting through (ἄπορα)? Look at the size of the rivers.... You wouldn't even be able to get across some of these rivers at all, without us to ferry you (διαπορεύοιμεν). (2.5.18; cf. further πόρους 2.5.20 and ἀπόρων 2.5.21)[18]

The strong verbal repetition here itself serves as a stylistic figuration of Persian abundance.

The spatial terms used to describe the Greeks' desperate position at the beginning of book 3 reappear in the account of the retreat. The march to the sea is marked by repeated returns to ἀπορία. Caught between the Tigris to their left and mountains to their right, the Greeks were in "great difficulty" (πολλὴ ἀπορία, 3.5.7; cf. ἀπορουμένοις 3.5.8). A way out of

17 Rood 2010b: 58. Cf. Hartog 1988: 57–60 on the use of πόρος/ἀπορία in Herodotus' Scythian narrative; also, for the *Anabasis*, cf. Purves 2010: 187–88, though her broader picture of the degree to which the Greeks seem lost in the interior is rather exaggerated. For the verbal play on πόρος/ἀπορία, cf. Eur. *IT* 897 ἀπόρων πόρον ἐξανύσας (Hermann's emendation of the MS πόρον ἄπορον; he also conjectured πόρον εὔπορον); Ar. *Frogs* 1465; also n. 21.

18 Cf. Cyrus' insistence in his exhortation to the Greek troops that he was "not short of barbarian troops" (οὐκ ἀνθρώπων ἀπορῶν βαρβάρων, 1.7.3). Contrast the use of ἀπορήσας of the Persian king at Isoc. 4.147, arguing that the fact that the king resorted to treachery and sacrilege was evidence of the weakness of his own forces.

their ἀπορία beckons when they are informed that once they were in Armenia, "they could easily (εὔπορον) go wherever they wanted" (3.5.17).[19] But when the Greeks discover that the River Centrites blocks their way out of the Carduchian land, they are "in great difficulty" (ἐν πολλῇ ἀπορίᾳ) (4.3.8), as they were after the arrest of the five generals (3.1.2). As we shall see, it is the character "Xenophon" who leads the Greeks out of ἀπορία thanks to his grasp of the tactical adjustments required in difficult terrain and to his ability to inspire the troops with his own vision of space.

The same conception of space is found even after the Greeks have returned to the sea. After the Arcadians have broken away from the rest of the army, they are cut off by Bithynian Thracians on a hill (6.3.6–8, cf. περικεκυκλωμένοι 6.3.11), and there was "great difficulty" (ἀπορία πολλή, 6.3.9). Later, when a contingent moves out to bury some troops killed in an earlier foray for supplies, "Xenophon" draws on their experience of the obstacles they have overcome and those that still face them on their return to explain to his men why they must cross a wooded ravine:

> I'm surprised that anyone would consider this ravine more formidable than the rest of the countryside we've just marched through. How is the plain to be crossed unless we defeat their cavalry? How are we to get back over the mountains with so many peltasts on our heels? And if we do get back safely to the sea, look at the size of the ravine that is the Euxine.... (6.5.19–20)

With plains and mountains to cross, they face a miniature version of their original return to the sea. "Xenophon"'s rhetoric

19 Ambler 2008: 267 n. 22 is tempted by ἄπορον, a manuscript variant: "This reading is in harmony with the difficulties the Greeks continue to face even after leaving Carduchia, but it also suggests that the generals do not report this bad news to the troops in general." It is easier to suppose a scribal slip: Xenophon is explaining the generals' decision, not revealing how they justified that decision to the troops; and the later disappointment is itself anticipated by the repetition of ἔφασαν ("they said") after εὔπορον.

successfully relates the troops' immediate experience of space (the ravine ahead) to a grander narrative of the overcoming of spatial obstacles.

"Xenophon" also foresees further threats to the army's freedom of movement should they alienate the Spartans, the dominant power in Greece. Once the troops have crossed the ravine, he encourages them with the thought that they are "at the doors of Greece" (6.5.23: ἐπὶ ταῖς θύραις τῆς Ἑλλάδος). Though part of a proud appeal to the memory of their achievements, "Xenophon"'s expression (which is not paralleled in extant Greek literature[20]) recalls earlier descriptions of the troops as "at the doors of the king" (2.4.4, 3.1.2).[21] Just as earlier their proximity to the king was a cause for alarm, so now their proximity to Greece proves to be dangerous owing to the Spartan presence. Soon "Xenophon" warns the troops that the Spartans may "shut us out" from the cities and make it hard "to stay or to sail away (καὶ μένειν καὶ ἀποπλεῖν)" (6.6.12–13) – just as during the mutiny at Tarsus the army had been warned of "the difficulty (ἀπορία) of either staying or leaving (καὶ μένειν καὶ ἀπιέναι) without Cyrus' consent" (1.3.13).[22] On the way out they were at the mercy of Cyrus. Now they are at the mercy of Spartans who have the power "to do whatever they like" (6.6.12) – a power generally associated with tyrants. These problems in due course materialize after they have crossed the Bosporus (e.g., 7.2.15, 7.3.3; cf. the summary at 7.6.24–26).

It is the words of "Xenophon" the character that provide the narrative resolution to a theme that Xenophon the narrator

20 Cf. Hansen and Nielsen 2004: 151 for the phrase as a demarcation of Greece (they compare Callisthenes, *FGrHist* 124 F 5.13, Byzantium as "the arm-pit of Greece").

21 Cf. Rood 2004: 319–20. I take this phrase primarily to highlight a parallel between two sources of danger rather than to configure Greece itself as an internal (Spartan-dominated) space.

22 The difficulty of staying or leaving (καὶ μένειν καὶ ἀπιέναι) also recurs (again in conjunction with an ἀπορία word) at 5.2.23, of a difficult retreat after an attack on a Drilian stronghold – another miniature version of the whole retreat, this time effected through divine help (cf. n. 28); cf. also 5.6.20, 7.6.44, 7.7.51 for the pairing of μένειν and ἀπιέναι.

has set up. When the Greeks finally agree to join the Spartans, "Xenophon" offers a positive gloss on their final position in a speech where he defends his own handling of events in Thrace: "you can now travel in safety wherever you choose, by land and sea.... your difficulties have substantially vanished" (7.6.37: literally, "you have much εὐπορία"). And yet this positive gloss is threatened by an intratextual echo. As we have seen, they had earlier been promised that once they crossed the Carduchian mountains and arrived in Armenia "they could easily (εὔπορον) go wherever they wanted" (3.5.17). That promise, it turns out, had been a considerable exaggeration. So too the closing cadence to the narrative – as most of the remaining soldiers sign up to serve the Spartans in a new venture into Asia – underlines the illusoriness of their temporary εὐπορία.[23]

We have seen, then, that Xenophon structures his presentation of the Greeks' engagement with the spaces of the Persian Empire by evoking broader Greek spatial conceptions and by a complex web of internal allusions. The language used for the Greeks' perceptions of their position and "Xenophon"'s own interpretation of his dream picks up a number of earlier passages where the vocabulary of ἀπορία and encirclement appears focalized both through the narrator and through characters. This verbal network further emphasizes the desperate position of the Greeks, showing how their worst fears are realized after the seizure of the generals. It also prepares for the way the Greeks' escape from their difficulties is described: the march back to the Greek world is marked by a frequent return to the difficulties that had surrounded the Greeks in Mesopotamia, with the repetition of the ἀπορία motif cutting across the narrative's forward momentum. We turn now to examine the

23 Xenophon's narrative of a thwarted spatial trajectory from ἀπορία to εὐπορία may also be read alongside the philosophical narrative of Eros, child of Poros and Penia (Poverty), at Pl. *Smp.* 203b1–204c6 (note εὐπορία words at 203e2, 204b6, 204e6, and ἀπορία words at 203b8, 203e4, 204b7; the roots are picked up at other points in the dialogue too); cf. also Aristarchus, *TGrF 1* 14 frag. 2: Eros "makes τὸν ἄπορον find πόρον."

similar narrative artistry displayed in Xenophon's account of how the Greeks do manage to escape – where it is "Xenophon" himself who is the beneficiary.

Landscape and Strategic Space

In his speech encouraging the army about their prospects for survival after the seizure of the generals, "Xenophon" argues that "we've seen with our own eyes how the Lycaonians have seized fortresses in the plains and cultivate land which belongs to the Persians" (3.2.23). This speech gives more detail about the landscape of Lycaonia than the earlier narrative, where the army marches "through Lycaonia, where Cyrus gave the Greeks permission to turn to plunder, since it was hostile territory" (1.2.19). Xenophon's main focus in the earlier narrative is on the fact that Cyrus has now crossed the boundary of the area allotted to him by his father; there is no overt hint that the Lycaonians are unruly. Here, by contrast, "Xenophon"'s appeal to something that all the troops have supposedly seen, but that the narrator has not mentioned, presents him as distinctive in his ability to draw a practical lesson from geography.

A similar spatial privilege is given to "Xenophon" in narrative descriptions of landscape during the retreat. These descriptions are themselves part of a narrative progression. In the account of the march up-country, the narrator offers a number of generally brief and formulaic descriptions of cities and rivers at which the army arrives. The only details generally given about the sections of the march in between stops are the numbers of stages and parasangs, which are mentioned in a highly formulaic phrase that reflects Cyrus' steady progress toward his goal; it is only for the Cilician plain (1.2.22) and the Arabian desert (1.5.1–2) that, exceptionally, some description of landscape is offered. During the retreat, the narrator continues to describe some places where the army stops, notably the ruined cities of Larisa and Mespila (3.4.7–9, 3.4.10–11); these passages tend to occur in sections where the army makes steady progress, as is signaled by the return of the stages/parasangs formula used for the march up-country. Descriptions of landscape, by contrast, are most detailed in sections where the Greeks face military obstacles that prevent them making that sort of steady progress – and they are often first given not directly by the narrator, but as seen by the character "Xenophon." The narrator's distribution of information underlines "Xenophon"'s distinctive spatial understanding: typically, "Xenophon" sees more than the other generals, and it is this capacity to see more that leads the Greeks to safety.

The distinctive vision attributed to "Xenophon" is itself the product of a development in the presentation of the march. During the army meeting following the capture of the generals, "Xenophon" had proposed that the army should adopt a square formation (3.2.36). In the early stages of their retreat, however, "the Greeks recognized" (3.4.19) the disadvantages of this formation whenever the road narrowed in mountains or for a bridge: "for it is necessary, whenever the sides of the square converge, either because the road narrows or mountains or a bridge make it necessary, that some hoplites are squeezed out . . . ; and when the sides diverge again, the hoplites who have previously been squeezed out are necessarily out of position" (3.4.19–20). Here the necessity imposed by the constraints of space matches a necessary human response. A shift occurs when Xenophon describes how the Greeks found a solution to these problems: it is now "the generals" who recognized the problems (3.4.21) and made plans for smaller mobile units (3.4.21–22) whose successful operations are described with the same order of exposition and the same language used in outlining the problems ("whenever the sides converged . . . whenever the flanks diverged . . . ").

In the next section of the march, Xenophon still focuses on the perceptions of the Greeks as a whole rather than on "Xenophon" in particular. He describes how, despite their adjustments, the Greeks continued to suffer from the Persian cavalry during the course of four days' marching along the Tigris. A change of terrain on the fifth day is then focalized through the Greeks: "they saw a palace of some kind with a number of villages grouped around it. The road to this spot passed through high hills, which were the foothills of the mountain under which the villages were situated. . . . the Greeks were delighted to see the hills" (3.4.24). These hills are first described as the Greeks see them rather than through the pattern of arrival familiar from the earlier narrative. The Greeks' delight (ἄσμενοι), however, proves to be misguided: as they went down the first hill, they came under missile attack from the barbarians above but were unable to drive them off by pursuing back up the hill (3.4.25–27). The same thing happened with the next hill (3.4.28). They changed their plans for the third hill and arrived at the villages after marching with peltasts on the heights above and the rest of the troops on the road through the hills. Once more Xenophon's description of terrain points up how the Greeks successfully adapt in response to difficulties. On their descent to the plain, Tissaphernes caught up with them again, and "necessity taught" the Greeks to camp in the first village they saw (3.4.32). The Persians attacked the villages, but it was easier for the Greeks to defend from a set position than on the march, especially as the Persians would encamp more than 60 stades away (3.4.33–34).

Xenophon then shifts his focus to the visual understanding of landscape shown by the Greeks' generals, the Spartan Chirisophus (who is leading the vanguard) and "Xenophon" himself (who is positioned at the rear). He first describes how the Greeks got ahead of the Persians once again, but on the fourth day the Persians went ahead and "occupied a position above the Greeks' intended route, on a spur of the mountain overlooking the way down to the plain" (3.4.37: καταλαμβάνουσι χωρίον ὑπερδέξιον ... ἀκρωνυχίαν ὄρους). Then Chirisophus at the front "saw that the spur was already occupied" (3.4.38: ἑώρα ... προκατειλημμένην τὴν ἀκρωνυχίαν) and ordered "Xenophon" in the rear to bring the peltasts up. "Xenophon" duly rode up, but without taking the peltasts, as he could see Tissaphernes approaching from behind. When he asked Chirisophus why he had summoned him, Chirisophus appealed to visual evidence: "Just look – the high ground overlooking our way down has already been occupied. We can't get past unless we clear them off it" (3.4.39). Chirisophus here uses a conventional Greek opposition between level and high ground. "Xenophon"'s distinctive grasp of space is shown by his ability to differentiate:

> Xenophon saw that the peak of the mountain was right above their own army, and that there was a way to approach the enemy's position from the peak. "Chirisophus," he said, "the best plan would be for us to make our way as quickly as possible to the top of the mountain. If we occupy that, it will be impossible for the men overlooking the road to stay where they are." (3.4.41)

The aspects of the landscape that offer a solution to the Greeks' strategic problems – the accessibility of the mountain peak – are described first as "Xenophon" perceives them. His vision trumps Chirisophus' because he can see that not all "high ground" is alike.

Chirisophus' grasp of space is again surpassed by "Xenophon"'s during the journey through the Carduchian mountains. Xenophon describes how the Greeks' rear suffered heavily from missiles from behind, forcing them to chase the enemy off. "Xenophon" would ask for a halt at such points. Once, however, Chirisiphus failed to stop, passing word to "Xenophon" to keep up with him. As soon as the opportunity arose, "Xenophon" went up to complain. Chirisophus then gave his reply: "Just look at the mountains ... and see how impassable they all are. There's only one road – a steep one, as you can see – and you can see that there's a horde of people on it. They've occupied the pass ... The guides we have with us say there's no other road" (4.1.20). Here the narrator, who has been focussing on events in the rear, reveals what has been happening ahead as Chirisophus describes the Greeks' position. His technique seems to suggest that Chirisophus is finally making appropriate use of his visual grasp of landscape. But that impression is deceptive: "Xenophon" surpasses him once more – this time by producing captives of his own who reveal an alternative route (4.1.24–25).[24]

"Xenophon" conquers spatial difficulties through a more unusual type of vision after the Greeks have crossed the Carduchian mountains. The narrator first describes how "they bivouacked among the villages overlooking the plain of the Centrites river ... glad to able to see the plain; the river was six or seven stades away from the Carduchians' mountains" (4.3.1). As earlier, the Greeks' pleasure (signaled by ἄσμενοι) proves to be premature: the next morning, "they saw armed horsemen on the far side of the river to stop them crossing, and foot soldiers lined up on the banks above the horsemen .... The banks on which these troops were arrayed were three or four plethra beyond the river, and there was only one visible road heading inland, which seemed to be man-made" (4.3.3–5). The narrator keeps up this focus on the vision of all the Greeks as he describes their "great despair (πολλὴ ἀθυμία) ... as they saw the difficulty of crossing (δυσπορίαν) the river, as they saw in front of them troops whose purpose was to stop them crossing, and as they saw behind them the Carduchians who would attack them as they were crossing" (4.3.7).[25] At this crisis (underlined by a return of the ἀπορία motif, for which see above), "Xenophon" "saw a dream" (4.3.8) – and the meaning of the dream was revealed when two young soldiers accidentally discovered a crossing-point where the Greeks would not be opposed by the cavalry.[26] By describing the dream,

[24] An account of how these men were captured is first given by "Xenophon" in his conversation with Chirisophus (4.1.22); this narrative delay further shows up "Xenophon"'s foresight.

[25] Note the triple use of ὁρῶσι, which is then matched by the anaphora of ὁρῶντες used with the barbarians as subjects twice at 4.3.21 and again twice at 4.3.23.

[26] In this case the narrator's precise spatial description underlines the fortuitous nature of the discovery of the crossing: "they had been out collecting kindling for a fire, when they had seen on the far side of river, among some rocks that came down to the water's edge, an old man, a woman, and some young girls putting what looked like bags of clothing inside a hollowed-out rock. They took a look and decided that it was safe to cross, because at that point the

Xenophon seems to appropriate for himself the fortunate discovery of the crossing.

The opposition between Chirisophus and "Xenophon" comes once more to the fore when the army finds the Chalybians, the Taochians, and the Phasianians opposing them at a pass from a plain into a new mountain range (4.6.5). Chirisophus stopped "when he saw the enemy at the pass" thirty stades away, "so that the Greeks would not get too close to them while still marching in a column" (4.6.6). He then summoned the officers and told them that "the enemy, as you can see, control the mountain pass" (4.6.7). "Xenophon" himself proceeds to offer a much more precise description of the mountain in a speech:

> We presumably want to get across the mountains with as little trouble as possible ... Now, the mountain range is, as far as we can tell from here, more than sixty stades long, but we can see that the only place they are guarding against us is the actual road. So it would be much better for us to try surreptitiously to steal a part of the mountain that is unoccupied. (4.6.10–11)

He then reveals that the rear has captured some guides (4.6.17) – a fact that the narrator has not mentioned earlier – before articulating an intellectual lesson in the military use of landscape: if the Greeks seize part of the mountain, "Xenophon" does not expect that the enemy will fight "on equal terms" (ἐν τῷ ὁμοίῳ), since "they're not prepared to come down and face us on level terms (εἰς τὸ ἴσον) even now" (4.6.18). He also draws an elaborate parallelism as he argues that marching without fighting uphill, or on rough ground, or at night, is easier than marching while fighting downhill, or over level ground, or by day (4.6.12). "Xenophon" offers an intellectual transformation of landscape: fighting against the enemy on the mountainside is equivalent to fighting on a plain; no less than the level plain,

terrain was unsuitable for the enemy cavalry;" and then after wading over "without getting their genitals wet" they stole the clothes (4.3.11–12). Cf. Hdt. 1.84 for a similarly fortuitous discovery of a point of access to the citadel at Sardis.

fighting on the hills can be conceived as fighting on level terms. His spatial insights are in due course confirmed by the narrative: the Greeks seize the mountain at night and advance the next day both along the path and through the high ground (κατὰ τὴν ὁδόν ... κατὰ τὰ ἄκρα, 4.6.23), and the barbarians at the pass flee when the Greek upper contingent is victorious.

A final instance of the narrator's manipulation of spatial information to highlight "Xenophon"'s superior insight during the march to the sea appears when the Greeks arrive at the land of the Taochians. The narrator notes that the Greeks were short of supplies because the Taochians had removed them to the strongholds where they lived. Chirisophus then made some small units attack one of the strongholds, which is described as "perched on top of precipitous cliffs, so the Greeks could not ... attack it all together" (4.7.2). When the generals consult, Chirisophus explains to "Xenophon" why the attacks are not succeeding: "There's only one way in – the one you can see – and whenever we try to go along it, they roll rocks down on us from that cliff overlooking the road" (4.7.4). In reply, "Xenophon" gives more precise details: "as you too can see, although the distance we have to cross while under attack is almost one and a half plethra, about a plethron of the ground has pine trees growing on it, with wide spaces between the trees" (4.7.6). He then suggests that the men should take cover behind the trees while advancing. Chirisophus objects that the enemy will roll stones down at them, but "Xenophon" sees this as a positive advantage (as they will use up the stones faster). When his plan is put into action, the narrator reveals how little cover there actually was: "some men ... took up positions before the shelter of the trees, because there was room for only one company of men to stand among the trees" (4.7.9). So here at least "Xenophon"'s perceptions are not quite accurate. But his plan still leads to the capture of the stronghold, and the army is able to proceed on its way toward the coast.[27]

[27] There are further obstacles on the way to the sea – a ravine on the borders of Macronian territory (4.8.2) and a mountain on the borders of Colchian territory (4.8.9) – but here the descriptions (with

After the Greeks' arrival at the Black Sea, Xenophon generally pays attention to landscape in his accounts of their incursions into the interior, not when he is describing their movements directly along the coast itself. Rather than exploring further the treatment of military space in Xenophon's account of these incursions,[28] I will discuss the two main exceptions – the description of Calpe Harbor and the proleptic account of the estate that "Xenophon" buys for Artemis at Scillus in the Peloponnese.

The description of Calpe Harbor appears when the different units of the army have re-united at a beach below a headland:

> This place, which is called Calpe Harbor, is in the part of Thrace which is in Asia. This part of Thrace, which is on the right as one sails into the Euxine Sea, starts at the mouth of the Euxine and goes up to Heraclea. For a trireme propelled by oars it is a very long day's journey from Byzantium to Heraclea, and there is no other friendly or Greek town between these two places, only Bithynian Thracians, whose treatment of any Greeks who fall into their hands as a result of shipwreck or whatever is said to be horrendous. Calpe Harbor is situated at the halfway of the voyage from Byzantium to Heraclea or vice versa . . . (6.4.1–3)

Xenophon goes on to describe how there is a headland above a sheer cliff that is connected by a narrow neck to the mainland,

imperfect tenses) are focalized not through "Xenophon" but implicitly through the army. In the first case the difficulty is solved thanks to a Macronian peltast who recognizes the language, in the second thanks to "Xenophon"'s strategic acumen.

28 See above on 6.5.19–20. Note also 5.1.14, where "Xenophon" secretly overrides the army's reluctance to prepare a route by land by persuading the cities along the coast to help make the roads "serviceable" (εὔποροι); and 5.2.3–5, where the difficult terrain in front of a town is implicitly focalized first through the army as a whole, then through the light-armed troops who lead the attack. As in the Colchian mountains, it is here "Xenophon"'s strategic acumen rather than his distinctive vision of space that proves decisive – though even so divine help is required to ensure a safe retreat (5.2.24 – cf. n. 22 above).

that is itself "large enough to accommodate ten thousand people" (6.4.3), and that also has a harbor with a beach and a spring. In the vicinity, he adds, there is plentiful timber as well as good soil for some distance into the mainland. The whole inland district is "fine and plentiful," with "many inhabited villages": "the soil produces barley, wheat, all kinds of legumes, millet, sesame, a good number of figs, plenty of grapes which make a sweet wine, and so on and so forth – everything except olives" (6.4.6).

The description of Calpe Harbor differs from the passages discussed above in that it is focalized through the narrator rather than through "Xenophon." Indeed, it has many of the attributes of the earlier narratorial account of the Cilician plain (1.2.22), where the present tense is also used for crops. The narrator makes it clear, however, that the troops too perceived the site's advantages: "they demonstrated no desire to do so at the place which might have become a town. In fact, they thought that it had taken scheming, on the part of those who wanted to found a city there, even to come to this place" (6.4.7). The narrator here alludes to "Xenophon"'s earlier perception of the advantages of founding a new city along the shores of the Black Sea and to the hostile Greek response to this idea (5.6.15–7.11). As has often been noted, the description of Calpe Harbor is designed to highlight the potential of the site for a new settlement. The soldiers are not keen to settle along the shores of the Black Sea, but their immediate perception of the site's potential nonetheless functions as a strong affirmation of their *polis* mentality.[29]

The soldiers' suspicion of "Xenophon"'s scheming underlines, at the same time, the growing internal tensions within the

[29] Larsen 2004: 477–78 suggests that Xenophon's account of Calpe Harbor closely fits the template for city foundations presented by Aristotle in the *Politics* (1325b33–27b15), though the connections seem slightly loose (in Aristotle, for instance, the desirability of a location right by the sea is a matter of debate); and note that Dillery 1995: 267 n. 71 identifies a number of utopian motifs (abundant springs, beautiful trees, abundance of fruit).

army. The following narrative brings out, moreover, that the Greeks' suspicions about plans for a new city blind them to the danger of remaining on the beach. As the army grows short of provisions, "Xenophon" advises that they should leave their baggage in a better defended spot on the headland, but the soldiers respond with indignation. The wisdom of that advice is shown when the Bithynians do threaten the army during the night. The following day the army moves to the headland and protects itself with a trench and palisade. Once again, then, "Xenophon" is shown to have a superior sense of strategic space. The other soldiers are blinded by their long-term fears to short-term strategic necessity.

That Xenophon's description of Calpe Harbor looks beyond the immediate needs of the Greek soldiers is suggested by the way he locates the site as "on the right as one sails into the Euxine Sea" rather than on the left as one sails out (the direction in which the Greek soldiers were traveling). The location of Calpe Harbor half way between Herakleia and Byzantium also suggests that founding a city would be a rational use of space, resembling in some ways the regularity of the (non-Greek) Mossynoecian pattern of settlement ("towns were located at intervals of more or less eighty stades, but the layout of the land, with its mountains and valleys, was such that the inhabitants of one town could hear people shouting to them from the next one" 5.4.31), or the Thracian system of dividing shipwrecked goods ("The Thracians who live in this region set up marker-stones and each group of them plunders the wrecks that fall within their plot" 7.5.13). Yet Xenophon's account of the response in the neighboring cities to the prospect of a settlement makes clear that a new city could flourish only with the cooperation of the existing Greek cities along the coast. "Xenophon" once more stands in isolation. But now he is isolated not in his ability to read spatial signs, but in his ambitions for Greek expansion within the Black Sea. The other Greek troops are quick to read the site of Calpe Harbor as a likely site for a new city. But the tensions among the Greeks that reappear in this setting further underline the fragility of the Greek unity that had been all too briefly constructed in response to the adverse environment of Mesopotamia.

Xenophon's description of a place that "might have become a town" is matched in the closing three books by the description of the landscape of the sanctuary that "Xenophon" set up for Artemis on his return to Greece. This account appears when the Greeks have arrived at Cerasus. At this point their generals divide up among themselves a tithe for Apollo and Artemis from money realized through selling prisoners of war. Xenophon then explains how "Xenophon" dealt with his share of the money: he dedicated the portion for Apollo at Delphi while buying an estate for Artemis at Scillus in the Peloponnese, where he had been settled by the Spartans. He then offers an account of the estate: there was a river Selinus running through the estate containing fish and shellfish; the sacred land "contained meadows and tree-covered hills" with good fodder; around the temple there was "an orchard containing every kind of tree that produces edible fruit season by season;" and the local inhabitants would come together each year for a festival where they would enjoy this produce while the young men went hunting (5.3.7–13). In context, this idealized image of the setting for a new Greek festival involves a striking narrative displacement: a speech made by "Xenophon" later in the work warning the army against its growing indiscipline offers an analeptic narrative of some violent acts committed by some of the troops at Cerasus against the local inhabitants (5.7.17–31). Read against the text as a whole, the account of the festival at Scillus stands out as the only place where Xenophon brings out strongly the emotional attachment of a particular place for the local population. This evocation of local attachment to place is often seen as nostalgic – perhaps a sign that Xenophon was writing after his expulsion from Scillus in the aftermath of Sparta's defeat at Leuctra.[30] Far from being nostalgically picturesque, however, the precise spatial

[30] Cf. how modern humanist home-centered notions of "place" have been criticized as nostalgic and gendered.

details Xenophon offers about the site all bolster "Xenophon"'s implied claim to have fulfilled his duties to Artemis: he stresses that a tithe is offered to the goddess from the abundant annual yield from the Scillus sanctuary, and also that the sanctuary itself is a small-scale replica of her sanctuary at Ephesus (there is also a river Selinus at Ephesus with fish in it, and the goddess' temple and statue are modeled on those at Ephesus). "Xenophon" gratifies the goddess with the same spatial vision that he showed in leading the soldiers back from Mesopotamia, while defusing any hint of excessive pride by insisting that the choice of the site was approved by an oracular response given by "the god" (presumably either Apollo at Delphi or Zeus at Olympia).[31] He creates the illusion of a sense of place that is – at least temporarily – immune from the divisions that repeatedly surface in the later stages of the Greeks' retreat.

LANDSCAPE AND IDENTITY

This chapter has explored Xenophon's depiction of space and landscape from two interrelated angles. I have suggested that Xenophon does often view landscape through a narrow strategic lens and as a stage for the Greeks to display endurance and adaptability in surmounting their difficulties. But I have also highlighted a sequence of descriptions where the character "Xenophon" is seen displaying a superior vision of the military possibilities of mountainous terrain – a capacity to read physical surroundings that sets him apart from the other generals and insures the Greeks' safety. While earlier historians had sometimes attributed specific strategic insight to particular individuals,[32] those insights were not the result of the sort of impromptu vision displayed by "Xenophon" in the retreat.

[31] The form of the question is left open: it was probably a request to approve the site that "Xenophon" had already chosen (cf. Purves 2003: 92–96 for discussion).

[32] E.g., Hdt. 7.235: Demaratus offers Xerxes specific advice about Cythera; Thuc. 4.3: Demosthenes explains the advantages of Pylos to his skeptical colleagues.

Herodotus in particular is more concerned with clever ways of circumventing geographical features that are plain to all than with the distinctive spatial insight of individuals. The strategic mastery of "Xenophon" looks ahead rather to stereotypical portrayals of Alexander, constantly driven by a desire to overcome ἀπορία, and Polybius' stress on a swift grasp of spatial possibilities as a hallmark of generals such as Hannibal and Philip V.

"Xenophon"'s strategic vision of landscape itself unfolds against a backdrop of extreme narratorial evasiveness in the earlier stages of the march. Read in the light of later sections where "Xenophon"'s distinctive vision is evoked, the more frugal attention to landscape in the march up-country could suggest a deliberate narratorial tactic: Xenophon avoids descriptions with strong suggestions of autopsy to prepare the ground for the emergence of the master general "Xenophon." The general reticence of the march up-country facilitates the studied disassociation of the narrator from the character that is central to Xenophon's self-presentation.

Do Xenophon's landscape descriptions also contribute to his presentation of Greek cultural identity? To answer this question it may be helpful to return to this chapter's two epigraphs: Eric Newby's account of the sudden appearance of the famed explorer Wilfrid Thesiger at the end of his classic *A Short Walk in the Hindu Kush* (first published in 1958) and Xenophon's account of how the Greeks celebrate their arrival at the sea by holding athletic games on a rough mountainside chosen by a Spartan exile called Dracontius (4.8.26). Newby's famous account of his encounter with Thesiger evidently offers further demonstration of the importance of attending to narrative voice in descriptions of landscape: the anecdote forms a suitable close to his witty exercise in narratorial understatement. His first person narrative involves the creation of a narrative pact very different from that involved in working through Xenophon's portrayal of "Xenophon." At the same time, following Newby's report of Thesiger's fireside chat about the operations he has carried out on the locals ("Why, the other day I took out an eye. I enjoyed that."), the aphorism that closes the book ("God, you must be a couple of pansies") conjures up a vision of landscape that is deeply bound up with shifting cultural identities. On the one hand, there is Wilfrid Thesiger – a presence from the imperial past, matching through his own toughness ("as hard as nails") the tough mountain landscape ("like iron"); on the other, there is Newby himself – exponent of a different, more ironic and self-depreciating, mode of Englishness.

The laconic wit that Newby attributes to Thesiger is matched in Xenophon by the Spartan exile Dracontius. And just as Newby's anecdote parades side by side competing visions of Englishness, so too Xenophon's account of the games that Dracontius organizes on the slopes above Trapezus allows for further indirect reflection on how the Greeks' sense of identity has been shaped by their experiences. Athletic contests on a rough hillside rather than on a level plain (as in Greece itself) are suitable for men who have successfully adapted to the mountains of Kurdistan.[33] And yet it is only Dracontius who shows this immediate understanding. It takes this suitably laconic Spartan exile to get the point across to the generals – who presumably include "Xenophon" – in an aphorism that provides a retrospective insight into the spatial dynamics of Xenophon's narrative.

By taking a broader view of how landscape fits into the spatial patterning of Xenophon's work, then, we can see that the very alien quality of the landscape Xenophon describes draws out more strongly the distinctive cultural identity of the Greek soldiers. It is not that adaptability is expressly identified by Xenophon as a key component of Greekness: there are strong rhetorical appeals to Hellenic identity in the account of the retreat, but the distinctiveness of that identity is itself paradoxically threatened by the fact that the troops have to accommodate themselves to local variations of landscape in order to survive – and that they also have to take on the mantle

33 Purves 2010: 185–86 suggests rather that the unit of the stade used for the race marks a return to a measured Greek world.

of hostile invaders of innocent lands.[34] But, as the Spartan Dracontius intimates, the self-conscious cultural performance of athletic games to celebrate their return to the familiarity of a coastal environment ("The sea! The sea!") offers a complex comment on the link between landscape and identity. At the same time, at the very moment when the Greek soldiers have returned to the fringes of a Greek world that is now dominated by Sparta, the anecdote in which a Spartan is given control over space also looks forward to developments in the narrative of the march along the coast. The troops' emblematic Greekness is shown – and even enhanced – as much by their display of versatility and resolution in overcoming spatial difficulties as it is by their fragmentation once they have surmounted the obstacles that surrounded them.[35]

Bibliography

Agnew, J. 2005. "Space: Place." In *Spaces of Geographical Thought: Deconstructing Human Geography's Binaries*, ed. P. J. Cloke and R. J. Johnston. London. 81–96.

Ambler, W. 2008. *Xenophon: The Anabasis of Cyrus*. Ithaca.

Baslez, M.-F. 1995. "Fleuves et voies d'eau dans l'*Anabase*." In *Dans les pas des Dix Mille: Peuples et pays du Proche-Orient vus par un Grec*, ed. P. Briant. *Pallas* 43: 79–88.

Dillery, J. 1995. *Xenophon and the History of His Times*. London.

Dodds, E. R. 1960. *Euripides: Bacchae*. Oxford.

Fehling, D. 1989. *Herodotus and His 'Sources': Citation, Invention and Narrative Art*, trans. J. G. Howie. Leeds. (Germ. orig. 1971.)

Hall, E. 1987. "The Geography of Euripides' *Iphigeneia among the Taurians*." *AJP* 108: 427–33.

Hansen, M. H., and T. H. Nielsen. (eds.). 2004. *An Inventory of Archaic and Classical Poleis*. Oxford.

Hartog, F. 1988. *The Mirror of Herodotus: The Representation of the Other in the Writing of History*, trans. J. Lloyd. Berkeley, Los Angeles, and London. (Fr. orig. 1980.)

34 For the Greeks' adaptability in (non-military) domestic and sympotic contexts, see Roy 2007: 74–75; cf. also Rood 2010a: 153–55 for Xenophon's portrayal of the Greeks adopting non-Greek measures in their march to the Black Sea while also following the figurative path of the Persian invaders of Greece.

35 Cf. Purves 2010: 186.

Larsen, S. E. 2004. "Landscape, Identity, and War." *New Literary History* 35: 469–90.

Lee, J. W. I. 2007. *A Greek Army on the March: Soldiers and Survival in Xenophon's Anabasis*. Cambridge.

Massey, D. 1993. "Politics and Space/Time." In *Place and the Politics of Identity*, ed. M. Keith and S. Pile. London and New York. 141–61.

May, J. 1996. "Globalization and the Politics of Place." *Transactions of the Institute of British Geographers* 21: 194–215.

Newby, E. (1958) 2008. *A Short Walk in the Hindu Kush*. London.

Pelling, C. B. R. 2013. "Xenophon's and Caesar's Third-Person Narratives – Or Are They?" In *The Author's Voice in Classical and Late Antiquity*, ed. A. Marmodoro and J. Hill. Oxford. 39–76.

Purvis, A. 2003. *Singular Dedications: Founders and Innovators of Private Cults in Classical Greece*. New York.

Purves, A. 2010. *Space and Time in Ancient Greek Narrative*. Cambridge.

Rambaud, M. 1974. "L'espace dans le récit césarien." In *Littérature gréco-romaine et géographie historique: Mélanges offerts à Roger Dion*, ed., R. Chevallier. Paris. 111–29.

Riggsby, A. 2006. *Caesar in Gaul and Rome: War in Words*. Austin, TX.

Rood, T. 2004. "Panhellenism and Self-Presentation: Xenophon's Speeches." In *The Long March: Xenophon and the Ten Thousand*, ed. R. Lane Fox. New Haven, CT. 305–29.

2006. "Herodotus and Foreign Lands." In *The Cambridge Companion to Herodotus*, ed. J. Marincola and C. Dewald. Cambridge. 290–305.

2007. "Xenophon." In *Time in Ancient Greek Narrative* (*Studies in Ancient Greek Narrative* 2), ed. I. J. F. de Jong and R. Nünlist. *Mnemosyne* Supplement 291: 147–63.

2010a. *American Anabasis: Xenophon and the Idea of America from the Mexican War to Iraq*. London and New York.

2010b. "Xenophon's Parasangs." *JHS* 130: 51–66.

2012. "A Delightful Retreat: Xenophon and the Picturesque." In *Xenophon: Ethical Principles and Historical Enquiry*, ed. F. Hobden and C. J. Tuplin. Leiden and Boston. 89–121.

Roy, J. 2007. "Xenophon's *Anabasis* as a Traveller's Memoir." In *Travel, Geography and Culture in Ancient Greece, Egypt and the Near East*, ed. J. Roy and C. E. P. Adams. Oxford. 66–77.

Waterfield, R. 2005. *Xenophon: The Expedition of Cyrus*. Oxford.

Wright, M. E. 2005. *Euripides' Escape-Tragedies: A Study of Helen, Andromeda, and Iphigenia among the Taurians*. Oxford.

3

IN THE BEDROOM: INTERIOR SPACE IN HERODOTUS' *HISTORIES*

Alex Purves

Xerxes' famous surveying of his troops from a hilltop at Abydos before crossing the Hellespont is paradigmatic of a certain view of space that is presented in Herodotus' *Histories* (7.44):

> When they arrived at Abydos, Xerxes wished to look upon his whole army. A throne of marble had been prepared beforehand for him there on the hill (the Abydonians had done this for the king who had ordered it in advance) and so there he sat, and looking down on the shore he gazed at his foot soldiers and his ships . . . [1]

In a manner that is similar to the extensive descriptions of territory and terrain that can be found in the ethnographic sections of Herodotus' work, Xerxes' view here accords with a geographical and cosmic perspective of space in early Greek poetics that is often associated with an outdoor or open view.[2] In this chapter I wish to examine how Herodotus offsets the extensive and geographical emphasis of his global project with a persistent interest in rooms and indoor spaces. These rooms take many forms, from the bedroom to the dining

1 Translations are my own.

2 The scholarship on this topic is extensive. See e.g., Romm 1992; Rood 2006 (see also Rood, Chapter 2 in this volume); I have discussed Herodotean exterior space, as well as the interior space of Xenophon's *Oeconomicus*, in Purves 2010: 118–58 and 196–233.

room, treasure house, or sanctuary. Plenty of very important things happen in them, yet as spaces they have been relatively understudied.[3]

In what follows, I pose the following questions about Herodotus' depiction of interior spaces as a counterpoint to those traditionally posed about exterior ones. Does the intimate space of the bedroom invite a different kind of reading (a "closer" reading, perhaps?) than spaces that are represented through views and landscapes? If the room is the invisible place from which we typically but often unconsciously situate ourselves when thinking about landscapes (since we sometimes see landscapes as or through windows, or as paintings hanging on walls), then what does it mean to turn the body around, away from that view, and to look back inside? Finally, if the landscape observer is typically imagined as both distanced from and owner of her view,[4] where does the observer of a room stand in relation to the scene she surveys? Inside or out? Peeking through or from behind the door, or moving confidently through a room of one's own with a full knowledge of its interior?

There has been much emphasis placed on landscape as "a way of seeing" and as a reflection of the bourgeois mastery of the eye over space (Cosgrove 1985), or the gazing subject's "detached authority" (Wylie 2006: 523). This means that the act of viewing a landscape (unlike the peek of an outsider into a bedroom) is neither furtive nor hidden; the one who looks at the landscape is the one who, by definition, has access to and often ownership of it.[5] In this chapter, I explore what happens when the furtive and transgressive peek of the outsider

3 One important exception is Sheila Murnaghan's analysis of secrecy and display in the Herodotean *oikos* (forthcoming), which has shaped this paper in significant ways, as have her comments on an earlier draft of this essay. I thank her for those comments and for sharing her paper with me.

4 As put succinctly by Bewes 2005: 87: "Landscape requires the presence of a perceiving subject; but it also requires that the observer stand outside it, in an attitude of aesthetic contemplation."

5 See further the papers in Mitchell 2002.

inwards, within the walls of a room, also triggers its own kind of ownership over a space within which the looker quite markedly does not belong. If the surveyor of a landscape is typically master of that space both before and after looking (with his gaze standing in for the unchanging manifestation of his power and interlocking naturally with what it looks at), the outsider's act of looking into a room marks an abrupt shift in the balance of power and events. In entering the bedroom, in other words, he leaves a mark on the space that will alter it permanently and can never be removed.

There are two different ways of thinking about the area of the room that are relevant to this discussion. Space can be considered in relation to how it is viewed (as with the landscape idea that I have suggested above); but, alternatively, it can also be understood in terms of how it is experienced. The two approaches focus on different capacities of the body; just looking, on the one hand, and moving (at the same time as looking and engaging the other senses), on the other.[6] While the landscape below Abydos is framed quite explicitly as the object of a gaze, in the stories from Herodotus that I present here we will see that it is impossible to *look at* a room without also physically *interacting* with it as a place. Here we find ourselves confronted with a frequently invoked contrast between what we might call embodied and disembodied space (often also referred to as the difference between "place" and "space").[7] For while "a site may be bodiless, . . . [involving] a disembodied overview, a survey" (Casey 1997: 204), "we get into place,

[6] For the phenomenological and kinesthetic aspects of space that emerge from the body moving through it, see especially Eckerman in this volume (23–35).

[7] Both Tuan and Casey argue for an opposition between "space" (abstract, global, framing, theoretical) and "place" (value-laden, felt, experienced, lived-in, embodied). This distinction can be useful, although at times it is also overstated and overly schematic (cf. the criticism of Soja 1996: 40, n.18, and the recent discussion in Thalmann 2011: 20–24); see also Lefebvre (1974) 1991: 39–46 on "representational spaces" and the Introduction to this volume (7–13).

move and stay there, *with our bodies*" (239, emphasis original).[8] The latter concept speaks in particular to the many ways in which human beings attach meaning to space (Tuan 1977: 5). That the entrance into a bedroom, for example, might activate feelings of transgression, illegitimacy, and desire in one person and comfort, familiarity, and possession in another, is a clear example of the fact that spaces mean different things to different people, depending on their familiarity with the environment, their status as insider or outsider, and the emotive or intellectual associations that the space might elicit. Second, the act of moving itself, whether into, out of, or through a space, contributes to the formation of it, just as the direction in which one chooses to move – whether forward, backward, or sideways – further adds a personal set of coordinates to space as it is experienced.[9]

Unlike a landscape, a room is a space that is traditionally not accessible to view; it is a personal, bounded space that it is usually not possible to enter without being seen by someone already inside.[10] Throughout the *Histories,* the outsider's act of looking within the room is repeatedly equated with desire, and one of the things I explore here is how that desire is activated specifically through the processes of entering and exiting. Part of what makes a room a room, I argue, is not only how it is constructed but also how individuals move into, out of, and through it. Once an outsider has moved inside a room he will find himself in an unstable and possibly dangerous situation, because he will need to find a position from which to look at

8 See also Tuan 1974: 10.

9 Tuan 1977: 12.

10 Between these two categories – landscape and room – there are of course many other versions of space, including those with indoor and outdoor aspects such as (in, say, Athens) the agora, theater, lawcourts (see further in Blanshard, Chapter 7 in this volume), or assembly. Here I focus on the landscape and the room as two opposing absolutes (I suggest, differently, ways in which one can be mapped onto the other in the *Odyssey* in Purves 2014). See n. 31 on the Greek/barbarian stereotyping of spaces as "outdoor" or "indoor."

things that do not belong to him. In Herodotus' rooms, it is hard to look without having designs on what you see, just as it is difficult to avoid being seen in the act of secretly looking.[11]

The room, moreover, with its simple interiority and its core association with the home, provides a space where the intimate and private connotations of feeling-through-the-body have particular resonance.[12] This is made most manifest in the bedroom, not only because it is a space that one has to move into in order to look at, but also inasmuch as it is a space where the highly personal acts of intercourse, sleeping, dressing, and undressing take place. If there is, in Tuan's words, an "intimate tie between place and self: the language used of the one is much the same as the language used of the other" (2004: 46), then in Herodotus the tie between the bedroom and the self is manifested, as I will go on to show, through both the practicalities and metaphors of dressing and undressing. Very often in Herodotus' work there is a strong correlation between the clothed body and the room, insofar as the room is revealed to view in a way that is often explicitly compared to the uncovering of the body of somebody inside it. This is related to the common Greek analogy between covering the body and being indoors, especially as it applies to women. As Llewellyn-Jones has shown, there is a clear association between the veil and the house, for example.[13] The fact, too, that the garment proves to be such an important metaphor in the key Herodotean story of Gyges and Candaules that we will begin by discussing (as

11 Solon is remarkable in being unaffected by his entrance into Croesus' treasure house (1.30). This is due in part to his wisdom and singular imperviousness to wealth, but also to the manner in which Croesus frames his room: he makes it too accessible, stripping any notion of secrecy or transgression from Solon's view. Note also that Solon is guided around it by servants.

12 See esp. Bachelard 1964.

13 Llewellyn-Jones 2003: 192–99, especially on the association between the material that covers a woman's body or face and the roof of the house (e.g., ἐπίβλημα [covering, veil] and ἐπιβλής [cross-beam in a roof], note also the late classical τεγίδιον [face-veil, or "little roof"], as well as on the analogy between doors and veils in the visual record (195–96). See further Cairns 2002.

summed up in the famous phrase "a woman takes off her αἰδώς along with her dress," 1.8.9) means, as scholars such as Ferrari and Cairns have demonstrated, that the symbolism of covering and clothing is already an important one in both the *Histories* and in Greek poetics at large.[14] What I want to do here is tease out how that network of images applies, specifically, to the notion of entering and exiting a room and the play or "texture" of space within it.[15] Finally, the fact that each of the rooms we will consider here also belongs to tyrants situates these spaces on an axis of possession, desire, and power, meaning that the acts of undressing or entering and exiting that take place in them become all the more charged.

I start with a close reading of Candaules' bedroom, which will serve as a model for my discussion of other Herodotean rooms to follow. As I move through the argument, I suggest that these rooms are important in terms of their architecture, access, and visibility, as well as for the stories that are told within them.

CANDAULES' BEDROOM

The scene that takes place in Candaules' bedroom at almost the very beginning of the *Histories* has been considered from many different angles, but let us begin with the way that Candaules specifically tells his bodyguard, Gyges, to view it (1.9.2–3):

> ἐγὼ γάρ σε ἐς τὸ οἴκημα ἐν τῷ κοιμώμεθα ὄπισθε τῆς ἀνοιγομένης θύρης στήσω· μετὰ δ' ἐμὲ ἐσελθόντα παρέσται καὶ ἡ γυνὴ ἡ ἐμὴ ἐς κοῖτον. κεῖται δὲ ἀγχοῦ τῆς ἐσόδου θρόνος· ἐπὶ τοῦτον τῶν ἱματίων κατὰ ἓν ἕκαστον ἐκδύνουσα θήσει καὶ κατ' ἡσυχίην πολλὴν παρέξει τοι θεήσασθαι. ἐπεὰν δὲ ἀπὸ τοῦ θρόνου στίχῃ ἐπὶ τὴν ευνὴν κατὰ νώτου τε αὐτῆς γένῃ, σοὶ μελέτω τὸ ἐνθεῦτεν ὅκως μή σε ὄψεται ἰόντα διὰ θυρέων.

14 Ferrari 1990; Cairns 1996, 2012.

15 At the end of my discussion, I will make some suggestions as to the correlation between layers of clothing and the materiality of space.

> I will place you in the room in which we sleep, behind the open door. After I have entered my wife will come to bed. A chair is situated near to the entranceway. On this chair she will place her clothes as she takes them off, one by one, and you will be able to watch at your leisure. But when she moves away from the chair and toward the bed and you are behind her back, then take care that she does not see you as you go out through the doors.

The plan, although it appears to be carefully constructed, for some unstated reason does not work. As the text continues we see Gyges agreeing reluctantly to the king's suggestion and attempting to follow it through (1.10.1–2):

> ὁ μὲν δὴ ὡς οὐκ ἐδύνατο διαφυγεῖν, ἦν ἕτοιμος· ὁ δὲ Κανδαύλης, ἐπεὶ ἐδόκεε ὥρη τῆς κοίτης εἶναι, ἤγαγε τὸν Γύγεα ἐς τὸ οἴκημα, καὶ μετὰ ταῦτα αὐτίκα παρῆν καὶ ἡ γυνή· ἐσελθοῦσαν δὲ καὶ τιθεῖσαν τὰ εἵματα ἐθηεῖτο ὁ Γύγης. ὡς δὲ κατὰ νώτου ἐγένετο ἰούσης τῆς γυναικὸς ἐς τὴν κοίτην, ὑπεκδὺς ἐχώρεε ἔξω. καὶ ἡ γυνὴ ἐπορᾷ μιν ἐξιόντα.

> The other, when he saw that he could not escape, was ready. So Candaules, when it was time for bed, led Gyges to the room, and quickly afterwards his wife entered. Gyges watched her come in and set down her clothes. When he was behind the back of the woman as she went towards the bed, he departed, slipping out. But the woman saw him going.

The repetition of two versions of the same scene at such close quarters in the text, with the events first previewed for the reader and then quickly afterward enacted, calls attention to the differences between the two passages. Herodotus, perhaps with a glance back to the Homeric precedent of close word-for-word scene repetition, reuses vocabulary and is precise in ensuring that the action that takes place mimics Candaules' stage directions. Yet the shift from the future to the past between the passages also compresses the scene, so that the entrance of the wife is narrated more succinctly, there is no longer mention of a chair in the room or of where it is situated, and Candaules' articulation of the taking off of items of clothing, one by one, and setting them on the chair becomes simply "her setting down (τιθέναι) the clothes." Although Candaules promised Gyges he could watch the scene "at leisure," that phrase is also dropped from the second passage, leading to a swifter impression of events.

The greatest number of differences between the two scenes, though, comes at the moment when Candaules' wife turns her back. Instead of going to bed like this: στίχῃ ἐπὶ τὴν ευνὴν, the queen now goes like this: ἰούσης τῆς γυναικὸς ἐς τὴν κοίτην, with changes in phraseology at every point along the line. Gyges, on the other hand, instead of exiting the room like this: ἰόντα διὰ θυρέων, now exits like this: ὑπεκδὺς ἐχώρεε ἔξω. Why would Herodotus place two scenes in such a symmetrical relationship, and yet work so hard, at the climax of both, to vary their vocabulary?

Candaules lays out the coordinates of his bedroom for Gyges with a careful use of spatial markers that plot the drama in a sequential string of prepositions, from ἐς (into [the room]) to ὄπισθε (behind [the door]), ἀγχοῦ (near [the door]), ἐπί (on [the chair]), ἀπό (from [the chair]), ἐπί (to [the bed]), κατά (behind [her back]), and finally διά (through [the doors]). But in the second passage Herodotus shifts these prepositional coordinates by rearranging and condensing them into the following sequence – ἐς (into [the room]), κατά (behind [her back], ἐς (toward [the bed]), and ἔξω (outside [the room]). In the second version of events, the prepositions focus only on a single idea; all that matters is that Gyges gets into and out of the room (ἐς to ἔξω) without being seen. There is no attention paid, unlike in the first version, to Gyges' desire. It is rather that Gyges understands that by entering the room he has entered a trap, from which he must try to get out as efficiently as possible. Yet implicit in the passage is the sense that his desire, just like the queen's shame, has been activated by the process of entering (and looking), and for that reason Gyges cannot exit as if nothing has happened in the room and as if nothing has changed.

Herodotus is careful in other respects, too, in the way that he switches the terms around between the two descriptions of the scene. In the transitional sentence between the two passages (1.10.1), we learn that Gyges will follow Candaules' instructions since he believes there is no escape from them (οὐκ ἐδύνατο διαφυγεῖν). Despite the fact that Candaules has just promised Gyges an escape route, he already seems to register that no exit is really possible for him.[16] Indeed, the δια- in διαφυγεῖν picks up and subverts Candalues' διὰ θυρέων from the sentence before (1.9.3). Next, in the midst of Herodotus' narration of what actually happens in the room, we find that the prepositional verb ἐκδύ[ν]ω (*to undress; to exit*) has been transferred from the wife's body to Gyges', as his slipping out of the door (ὑπεκδύς) in the second passage recalls her slipping out of her clothes (ἐκδύνουσα) in the first.[17] Now it becomes clear why the actual undressing of the wife was absent from the second version of events: Herodotus was saving the verb for Gyges. As if the room were playing tricks on its viewer and even on the reader, it unexpectedly presents us not so much with *Gyges* seeing the *queen undressing,* as promised in Candaules' version (ἐπὶ τοῦτον τῶν ἱματίων κατὰ ἓν ἕκαστον ἐκδύνουσα θήσει καὶ κατ' ἡσυχίην πολλὴν παρέξει τοι θεήσασθαι, 1.9.2), but with the *queen* seeing *Gyges exiting* (ὑπεκδὺς ἐχώρεε ἔξω. καὶ ἡ γυνὴ ἐπορᾷ μιν ἐξιόντα, 1.10.2).[18] Lest we forget that it is the very

16 Once he goes into the room, in other words, he will be lucky to make it out. His sense of foreboding mirrors Padel's reflections on what it means for tragic characters to enter the *skênê* on the ancient stage (Padel 1990, discussed later). Compare Rood's discussion of Xenophon's strategic ability to escape the encircling landscape of Asia Minor in the *Anabasis* (Chapter 2 in this volume).

17 LSJ s.v. ἐκδύω, ἐκδύνω; ἔκδυσις; ὑπεκδύομαι. Cf. δύω, δύνω.

18 There is an interesting transference between the two characters as vocabulary is passed back and forth between their roles in the first and second passages. The queen picks up much of the vocabulary originally assigned by Candaules to Gyges. She takes over the participle ἰών/ἰοῦσα when "going to the bed" (previously used of Gyges "going through the doors"). Similalry, ἡσυχίην is first used of Gyges watching the queen "at his leisure" κατ' ἡσυχίην (1.9.2), and then of the queen's "keeping quiet" even after she realizes she has been

allure of the wife's act of undressing that set the whole plot in motion, we might think of Gyges' double recourse to δύω in the phrase that he shouts out as soon as the king first proposes to him that he might observe his wife naked: ἅμα δὲ κιθῶνι ἐκδυομένῳ συνεκδύεται καὶ τὴν αἰδῶ γυνή ("a woman takes off her shame along with her dress," 1.8.3), where the *aidôs* is explicitly paralleled with an item of clothing.[19] The addition of the prefix συν- here further complicates this act of undressing, turning it into something more serious, more symbolic, and more consequential, much as the addition of the prefix ὑπ- does when added to ἐκδύ[ν]ω in Gyges' exit from the bedroom. The αἰδώς that should not be discarded acts, within the spatial imagery that the repeated use of ἐκδύ[ν]ω invites, as the last in a sequence of barriers (doors, clothes, shame) set up by society to prevent exposure of the woman before other men. These three barriers are fundamentally interleaved, to the extent that they all draw (metaphorically or otherwise) on the concept of covering and clothing.[20]

Once it has been performed in the space of the bedroom, the queen's disrobing becomes not just about her shame, especially insofar as it is compounded by a cultural code that specifies that even for a Lydian *man* to be seen naked is a great dishonor (1.10.3 and 1.8.3, above). It is also about the fact that Herodotus wishes to draw a connection between the taking on and off of clothes and the entering and exiting of rooms.[21]

spied on (ἡσυχίην εἶχε, 1.11.1). Finally, the verb ἀμβοάω/ἀναβοάω ("to cry out in astonishment") denotes the sound that is first articulated by Gyges in horror at Candaules' plan, and later suppressed by the queen when she sees that plan come to fruition (1.10.2; Flory 1987: 37).

19 Cairns 1996: 79: "the αἰδώς itself is a type of garment put on to conceal what lies underneath" (also his n. 14).

20 See further Ferrari 1990; Cairns 1996; Llewellyn-Jones 2003: 192–99. Worth noting in this context are the images on vase paintings where the act of holding a veil over the face is framed by the woman's position behind a half-open door (Llewellyn-Jones 2003: 195–96).

21 The verb ἐκδύ[ν]ω, (including its use here in the compound συνεκδύομαι) and its associated noun ἔκδυσις, occur throughout the *Histories* to describe the activities of undressing and exiting. So too

Both constitute central turning points in the Gyges story and together define Gyges' predicament in entering the bedroom of the queen. The putting on and taking off of clothes, which already has a rich metaphoric tradition in Greek poetics, here transfers onto the activities of entering and exiting the bedroom. In both contexts, the symbolic and erotic overtones of the uncovered body remain in place.

For Gyges, the room is a place of danger, from which his only hope is to escape. Yet it is also a place of safety for him, inasmuch as it is only in actually leaving the room that he puts himself at risk. His act of leaving is like a removing of his clothes, because it means emerging out of a protective shell and being exposed. Undressing thereby also stands as a metaphor for the exposure that occurs when someone is seen leaving a room, as it becomes known where he has been and therefore what he wants.[22]

If the two primary meanings of δύ[ν]ω are brought together in the story of Candaules and Gyges, to the extent that the taking on and off of clothes in the bedroom also reflects the act of entering and exiting that space (and vice versa), then we have to think for a minute about the significance of the door in the architecture of the room. I suggest that it is the presence of doors more than anything else (more than the walls,

do other forms of δύ[ν]ω, such as ἐνδύ[ν]ω, ἐσδυ[ν]ω, and ἀποδύ[ν]ω, ὑποδύ[ν]ω to mean entering and/or exiting, dressing and/or undressing. Other instances of ἐκδύ[ν]ω and related words in the *Histories* are: Histaeus promises Darius that he will not take off (ἐκδύ[ν]ω) his tunic after arriving in Ionia until he has made Sardo a tribute-paying subject (5.106); in the story of King Rhampsinitus and the Thief, the king is amazed to see that his gold has gone down, even though there is no obvious exit (ἔκδυσις) from his treasury (2.121γ, discussed later in the paper); vipers are described as seeking an exit (ἔκδυσις) from their mother's womb (3.109); the Greeks are predicted to have no exit strategy (ἔκδυσις, 8.100.3).

22 Just as the queen exposed her body when she was secretly seen disrobing by someone whom she did not see at first but did see afterward, so Gyges exposes his body when he is secretly seen exiting the room by someone whom he did not see at first but realized was watching afterward.

roof, or furniture) that differentiates rooms from other kinds of spaces.[23] The ability to open or close the door, to pass through it, to be denied entry at it, to hesitate on its threshold, to stand behind it, or to escape out of it, qualifies the relationship between the subject and the room. Bachelard is particularly good at underscoring the point that the relationship between a closed and open door is neither straightforward nor symmetrical (1964: 73):[24]

> In the domain of values, a key closes more often than it opens, whereas the door-knob opens more often than it closes. And the gesture of closing is always sharper, firmer and briefer than that of opening.

If we draw attention to a certain poetic or symbolic value that is attached to the opening or closing of doors, then we might return to the door that Candaules promised his bodyguard he would open for him and leave open for as long as it took for the queen to undress: ὄπισθε τῆς ἀνοιγομένης θύρης στήσω ("I will position you behind the open door"). The open door is more than just a way in for Gyges. Since it remains open (so that he can observe unobserved behind it) it also allows the reader to follow the story into the room.[25] Padel has discussed the symbolism of the half-open door in the context of fifth-century tragedy, in whose plots the most dramatic events take place inside and out of sight, through the *skênê* doors (1990: 357, emphasis original):

> On vase paintings from about 430 onward, they paint one or both leaves [of the *skênê* door] open and receding

23 In this light, it is particularly interesting to consider the metaphorical role of doors to designate the edge of the King's space in Xenophon's *Anabasis*, as pointed out by Rood (Chapter 2 in this volume). On the difference between inside and outside (and the mediation between the two), note Vernant's observations (1983: 127–75) on Hestia and Hermes as complementary spatial forces. Compare the ambiguity that pertains to city walls in Dougherty in this volume (Chapter 4).

24 See also his discussion at 222–24.

25 Travis 2000.

> inwards, foreshortened. Often there is a figure within, half-hidden behind the opened door... Whoever began it, surviving vase paintings of the tragic scene which choose a half-open door, with perhaps a figure half-hidden behind, to represent a tragedy, are a perfect emblem of tragic *feeling*. They express the tension between secrecy and revelation, the hidden and the manifest, which variously characterizes tragic texts. The door half-displays what is lying in wait, both in space (within the *skênê*) and in time, in the still-to-be-unfolded play.

As the Würzburg fragment mentioned by Padel shows, the figure situated just within the doors is visible to the spectator, allowing a preview of the action that will take place offstage and later be revealed in tableau when the *ekkuklêma* is rolled out. Gyges, as the spectator who watches an interior scene from precisely that position of being half-hidden behind the door, hesitates between subject and object, spectator and actor. As a viewer who is unwittingly looked at, he stands in an indeterminate place both on and behind the threshold, so that we are unsure whether he has really *entered* the scene or not. Here, Taplin's work on dramatic entrances and exits is especially helpful (1977: 7):

> At first sight it might seem obvious that by "entry" we mean the movement which brings an actor into the field of vision of the audience, and by "exit" the movement which takes him out of it...[but] we cannot possibly know at what moment the actor crossed the line between out of sight and in sight of the audience.

So too for Gyges, who tries to pretend that he is not in the room at all, as if there were only two actors in the scene rather than three. It is the moment at which he becomes visible that determines his participation in the events at play, but there will always be uncertainty about precisely when that moment occurs.

To think of the room as a "playing space" is of course misleading in one specific sense: the spectators could never watch events taking place within a room because they could not see within the tragic *skênê* (a point I will return to later). I am not, therefore, suggesting a one-to-one correspondence between Candaules' bedroom and the tragic stage, or that we should put ourselves in the position of theater audience members when reading the queen's disrobing. I do wish to suggest, however, that thinking about the idea of the door and the idea of the room through these tragic models is extremely helpful in illuminating the terms of the drama Herodotus has set in motion. Indeed the tragic dimensions of this particular plot have long been noted, not least because we know it also existed as a tragedy.[26]

The tragic stage had two primary sets of entrances and exits: the doors within the *skênê* building that lead, as in the Gyges story, to a room, and the side entrances and exits, known as the *eisodoi* or *exodoi*, which are the roads in and out that lead often to the country on one side and the city on the other.[27] Padel defines the difference between the two types of entrances thus (1990: 343, emphasis mine):

> The *eisodoi* lead out and beyond, to change, but the door leads within, to closed space. There is no way out. What happens in there is *the plot's trapped outcome* and dead end, the image of inevitability.

Ἔσοδος is provided by the *LSJ* as an opposite term for ἔκδυσις (*exit*). But the *e[i]sodos* entryway of the theater does not have a door. It is all the more interesting, therefore, that when

26 The 16-line fragment (P.Oxy 2382) of this "Gyges tragedy" is of uncertain date, with some believing that it was produced before the *Histories,* and some after (Lobel 1950 and Raubitschek 1955 both believe the Gyges tragedy served as a model for Herodotus). The fragment contains a speech of the queen's relaying what happened in her bedroom on the night of Gyges' first transgressive entry. See further Page 1951; Evans 1955; Flory 1987: 30–38; Benardete 1999: 11–16; Travis 2000 (who makes some interesting points on the differing depictions of space between the *Histories* and the tragedy).

27 The fourth, less common entrance and exit is the crane that brought actors (often portraying gods) into the playing area from the back of the stage. See further Taplin 1977: 443–51.

Candaules presented his overly-optimistic version of the bedroom plan to Gyges, he used this word ("a chair is situated near to the entryway (ἔσοδος)," 1.9.2), thereby presenting the way into the room as a straight and open entrance path and stripping it of the ambiguities that accompany doors.

We have already noted through our reading of the phrase οὐκ ἐδύνατο διαφυγεῖν ("he was not able to escape," 1.10.1) how Gyges views his situation before he is even led into the queen's chamber. He is right in his assessment, for even though he does escape from the room, his exit is not without consequences. The queen, summoning him the next day, gives him an explicit choice of "paths" to turn down: Νῦν τοι δυῶν ὁδῶν παρεουσέων, Γύγη, δίδωμι αἵρεσιν, ὁκοτέρην βούλεαι τραπέσθαι· ("Now there are two paths available to you, Gyges, and I give you a choice of which one you wish to turn down," 1.11.2). In doing so, she redirects "the plot's trapped outcome" from Gyges to Candaules, allowing Gyges the choice of a road that leads "out" but also in – to a bedroom that he is now invited not only to desire but also to possess. In order for this to happen, however, the scene must be replayed one further time, in exactly the same place (ἐκ τοῦ αὐτοῦ μὲν χωρίου), except now under the direction of the queen. This time Gyges will follow the queen into the room, once the king is already in bed, and hide again behind the door. From there, instead of secretly slipping out of the room, he will secretly slip *in* (ὑπεκδύς). His movement out from behind the door thus occurs as a kind of interior exit. It again exposes Gyges, this time revealing him as both the killer of the king and the new king as well (1.12.1–2):

> καί μιν ἐκείνη ἐγχειρίδιον δοῦσα κατακρύπτει ὑπὸ τὴν αὐτὴν θύρην. καὶ μετὰ ταῦτα ἀναπαυομένου Κανδαύλεω ὑπεκδύς τε καὶ ἀποκτείνας αὐτὸν ἔσχε καὶ τὴν γυναῖκα καὶ τὴν βασιληίην Γύγης...
>
> She, having given him a dagger, hid him behind the same door. Afterwards, when Candaules had fallen asleep, Gyges slipped in, killed the man, and took possession of both his wife and his kingdom.

In this final scene the plotting is more advanced, the secrecy more explicit (note κατακρύπτει, which accentuates the physical notion of hiding in the space, as well as the notion of a material covering), and the danger more heightened.[28] The ἐκδύνω that began, in Candaules' narration, as the wife's innocent undressing (and whose figurative potential was instigated by Gyges' equating of her αἰδώς with a robe) has now, with the addition of the surreptitious prefix ὑπ-, become the climax of a wholly tragic plot. Gyges' second emergence from behind the door prefigures his own future undressing within the space of the room and claiming of the king's intimate space as his own.

A story that has ramifications for the entire geography of Asia begins in the narrow space determined by the dark shadow of a door and its distance from a chair upon which clothes are laid in the private act of undressing.[29] Herodotus' grounding of the Gyges story within the bedroom affords us a specific perspective on the events of history. By placing his characters in the same room in three separate tellings, he uses its space to plot a micro-story of secrecy, desire, and exposure within the larger canvas of world history. The open door behind which Gyges hides, the bed toward which the queen walks, and the relationship between her act of undressing and Gyges' movement out from behind the door are all key elements in an area defined by the intimately closed and disclosing coordinates of door, chair, and bed.

I will now turn to look at other rooms connected with key moments in the maneuvering of power in the *Histories* and at other doors which open "at particularly stressed junctures" in the work.[30] What Murnaghan has identified as a dynamic of uncovering in relation to the woman's body and the secrets

28 Candaules devises (μηχανάομαι, 1.9.1), but the queen carefully plans for revenge (1.10.2).

29 As we learn immediately following the tale's conclusion, the events that take place in Candaules' bedroom have a direct causal link with the Persians' overthrow of Lydia five generations later.

30 The quotation is from Taplin 1977: 3, in relation to entrances and exits in tragedy.

of the *oikos* is also bound up, I argue, with a deviant form of entering and exiting the room in order to look at, and then possess, its contents. It is precisely because the inside of the room is *not supposed to be looked at,* because it should be, like the interior of the *skênê,* a shadowy space existing only in the imagination, that its intersection with the notion of landscape – and indeed all of the outdoor space that is described in the *Histories* – becomes so problematic.[31] In what follows, therefore, I will continue to focus on the acts of dressing and undressing that take place behind doors (closed or otherwise) and the consequences of Herodotus' uncovering of the room's spatial and symbolic dimensions.

Rhampsinitus' Rooms

In the tale of King Rhampsinitus and the thief, two men secretly enter Rhampsinitus' treasury in order to steal his gold.[32] The plot gets in motion, just as the Gyges story did, with a set of instructions laid out in spatial terms. For, on his deathbed, the builder of the king's treasury explains to his two sons how he secretly left a stone loose in one of its exterior walls (2.121α2):

> σαφέως δὲ αὐτοῖσι πάντα ἐξηγησάμενον τὰ περὶ τὴν ἐξαίρεσιν τοῦ λίθου δοῦναι τὰ μέτρα αὐτοῦ, λέγοντα ὡς ταῦτα διαφυλάσσοντες ταμίαι τῶν ⟨τοῦ⟩ βασιλέος χρημάτων ἔσονται.

> Having explained everything clearly to them about the removal of the stone he gave them its measurements,

31 It is no accident that these spaces are associated with non-Greeks, and usually with tyrants. The Greek normative practice (and this follows on from my discussion of the tragic *skênê*) is to focus the eye on exterior spaces. See further Rood (Chapter 2 in this volume), on Xenophon's presentation of the Persian king "contained within an internal space – unlike Greeks, who stereotypically undertake their public roles in the open air".

32 On the folktale aspects of this story, see West 2007. On its narrative structure, Munson 1993. I discuss the relationship between touch and contact in this scene in Purves 2013.

> saying that if they kept close this information they would become the guardians of the king's riches.

The brothers then find the stone, remove it, and are able to slip into the treasure house without the use of a door. By entering through this secret passage, the thieves do everything they can to avoid the notion of display, with the result that the room itself appears to be untouched (ἀσινές, 2.121γ3) by their presence (adhering to their father's instructions, they "guard carefully" against the exposure of their actions). The verb that is twice used to describe their entry, ἐσδύνω, (2.121β2.2, 6) is a compound of the same verb (δύ[ν]ω) used elsewhere in the *Histories* to describe the putting on of clothes.[33] Both actions (the thieves' entrance and the practice of dressing) share the same level of secrecy and a special familiarity with the space of the room.[34] Instead of a story that turns on a fateful act of exiting/undressing, therefore, we now have one whose premise is the careful guarding of secrets and a wholly invisible entering.

The thieves' mode of entry and exit differs dramatically from the king's, of course, who upon "opening the door was amazed to see the coffers of his treasures diminished, yet he was not able to blame anyone since the seals on the doors were preserved and the room was locked" (2.121β1). The king's role as an official opener of (his own) doors is paramount throughout

33 Cf. ἐνδύω (*to put on; to enter*) at 1.24 (Arion donning his performing outfit); 2.42; 2.81; 3.98; 4.146; 4.78; 6.125 (Alcmeon, discussed later); 7.15, 17 (Xerxes and Artabanos switching clothes, discussed later); 7.64; 9.22.

34 There are other stories in the *Histories* where we see illicit or unusual acts of dressing and undressing take place alongside subterfugitive or foreboding entrances into rooms that end up being associated with going to bed that are not discussed here. Alexander of Macedon dresses his men up as women, hides daggers in their cloaks, and sends them into the dining room to kill their lecherous Persian guests (5.20); Atossa uses the bedroom as a space to persuade Darius to invade Greece, after she has disrobed to show her breast to the doctor Democedes (he gives her medical advice in exchange for her promise to persuade her husband to start a war, 3.133).

this part of the story. Two and three times he opens the door (δὶς καὶ τρὶς ἀνοίξαντι, 2.121β4) to his treasury, always to be confronted with the same paradox: the objects in the space have changed in some way (his gold is missing; a headless man has been caught in his trap), yet "the room is undamaged and there is no sign of an exit (ἔκδυσις)" (2.121γ3–4). There has been no obvious uncovering in the room, no violation such as the undressing (ἐκδύνω) of the queen before strangers, no visible exiting and exposure (ἐκδύνω) of a potential usurper like Gyges, and yet the diminished amount of gold means that someone must have been inside.

The thieves' ability to confound the owner of the treasury by entering it improperly and exiting it invisibly shows a subtle wresting of control not only of the room's contents but also its spatial properties (eventually, like Gyges, one of them will become king and inhabit the palace). Unlike Gyges, however, again and again the thieves manage to exit without being exposed. King Rhampsinitus, standing in his own personal space, realizes that his room has become alien to him and that somewhere in its walls or doors there is a secret that he cannot uncover. Even after he sets a trap within which one of the thieves is caught, the brother takes the head of the corpse out with him, leaving a body that cannot be identified.

The room, in this scenario, once breached is no longer private but alien to its inhabitant. The emphasis on physical property (specifically, gold) points us toward a different kind of interior space, one that is less intimate than the bedroom but which highlights the notions of secrecy and ownership. The point will be best illuminated by sidestepping to two shorter anecdotes concerning treasure rooms of different sorts from the *Histories*. The first is a tale involving the entry into Croesus' treasure house which works in an almost completely opposite way, and the second involves the invited yet sacrilegious entry into the tomb of Nitocris, queen of Babylonia, in order to take the valuables she was buried with.

Croesus' Rooms

In book 6, we learn that Alcmeon pleased Croesus so much in Delphi that the king gives him the opportunity to go into his treasure house and come out with as much gold as he can carry. The youth prepares for this challenge in an extraordinarily straightforward manner. He puts on (ἐνδύς, 6.125.3) a huge tunic with a great fold in the front and huge boots. He then enters the king's treasure house and throws himself into heaps of gold dust, stuffing as much of it as he can into his clothing, boots, hair, and even mouth. When Alcmeon emerges from the room, looking "nothing like a man" with his cheeks puffed out and barely able to move in his boots, Croesus bursts into laughter. In fact, the king is so delighted by the sight of his guest-friend decked in gold that he doubles its amount for Alcmeon on the spot.

The story is interesting for what it tells us about different ways of going into and out of a king's room. Here, Alcmeon (who comes from a family twice described in this passage as "shining," λαμπροί,) engages in two stages of comical dressing – first with the overlarge clothes, then with the gold – that put his body on hyperbolic display, expanding and illuminating it. This childish, theatrical kind of dressing-up belongs to the outdoors and is safe and comical, establishing camaraderie between the two men that nevertheless reinforces their difference in status.[35] Alcmeon wants what Croesus has so much that he is *openly* prepared to play the fool to obtain it. At the same time, we see that the gold is so refined that – as the "stuff" that fills the room and sticks in Alcmeon's hair – it takes on something of the quality of neutral or simple space.

35 Cf. Kurke 1999: 143–46 on the ominous overtones of Alcmeon's costume; it is grotesque but also debased, Eastern, and feminizing, which she reads as a negative reflection on extra-polis aristocratic gift-giving.

NITOCRIS' ROOM

If the story of Alcmeon and his procuring of Croesus' gold is all about openness, invitation, and display, then the story of the Babylonian queen Nitocris shows a different aspect of the attempt to come out of a royal room with gold. Before her death, Nitocris arranged for her tomb to be placed over the most frequented gate in the city, and upon it she had inscribed the following words (1.187.2):

> τῶν τις ἐμεῦ ὕστερον γινομένων Βαβυλῶνος βασιλέων ἢν σπανίσῃ χρημάτων, ἀνοίξας τὸν τάφον λαβέτω ὁκόσα βούλεται χρήματα· μὴ μέντοι γε μὴ σπανίσας γε ἄλλως ἀνοίξῃ·
>
> If any of the kings of Babylon who come after me need money, then let him open my tomb and take as much as he needs. But may he not open my tomb if he is not in need.

Nitocris' inscription is a provocation, invitation, and prohibition all at once. By placing her tomb above the city gates, she denies entry to her burial chamber (the most personal of bedrooms)[36] as she facilitates entry to the public space of the city. Nitocris would seem to understand that the interior space of the room (especially a room belonging to a monarch) is finely balanced between public and private space in a way that is often difficult for tyrants to negotiate. Her repeated stress on the word ἀνοίγω (to open, as in a door)[37] stresses how easily the secret gold of her tomb might be exposed. It is Darius who finds himself unable to resist the inscription's lure; he breaks into the tomb to find nothing inside except the corpse and an approbation: "If you were not insatiably desirous of money and shamelessly greedy, you would not open the tombs of the

[36] On the tomb as a room, note the use of οἴκημα to describe, e.g., the underground burial vaults in the pyramids at 2.124.4. Cf. the Greeks' mythological pairing of the brothers Death and Sleep (Hes. *Th.* 212, 756–66; Hom. *Il.* 14.231, 16.682).

[37] Also *metaph.* lay open, unfold, disclose, LSJ *s.v.*

dead" (1.187.5). What Nitocris' trick illustrates so well is the strong combination of desire and secrecy that a closed, even sealed, room suggests. There is no dressing (-δύ[ν]ω) word in her story; instead ἀνοίγω doubles for the undressing and exposure that Darius' uncovering of her secret reveals.[38]

Each of the illicit intruders we have looked at so far enter out of a desire that is either forced upon them or suggested to them by someone else. Nitocris' ploy expertly combines the two spaces of treasure house and bedroom, wherein gold and women (presumably the two most desirable things of all) lie. As Murnaghan has argued, it is as if the owners of such rooms *need* their followers to believe there is something secret and desirable behind those doors, something worth breaking into the room for, in order to validate their royal power. There has to be, in other words, the idea of a closed secret – even if, as Nitocris so wonderfully reveals – that secret is no more than a private joke.

These anecdotes concerning Alcmeon and Nitocris show monarchs explicitly inviting entry for the purpose of free access to their goods. This is the same tactic that Rhampsinitus will try next in his attempt to catch the thief, with a turn away from the treasury and toward the bedroom and the body of his daughter. What he does is to set her up in a room (οἴκημα) whose doors explicitly defy convention by being held open to whoever wishes to enter through them.[39] The father tells

38 Darius' uncovering of the secret by opening the door points toward a kind of undressing inasmuch as it represents a stripping down to the truth. But it is also an undressing that suggests rape in its violation of the codes of intimacy and personal space, since the breaking in on a woman's body mirrors various perverse sexual acts carried out by tyrants, not the least of which is having intercourse with the corpses of queens. Cf. Hdt. 5.92 and Murnaghan forthcoming.

39 Rhampsinitus, as the father, explicitly overturns even more of the social and sexual codes of the bedroom, in inviting public access to his daughter's room, than Candaules did by allowing Gyges to enter the bedroom of his wife. The story of Rhampsinitus and the Thief's turn from personal space (signifying wealth) to intimate space (signifying privileged and exclusive access to the body) thereby immediately runs off course, circumventing the concept of room as intimate space even as it invites it.

his daughter she must sleep with each man who enters the room, but only after she has made that man first answer a question. The passage mirrors in interesting ways the Gyges and Candaules story and the use made by the king of a bedroom there (2.121ε2):

> τὴν θυγατέρα τὴν ἑωυτοῦ κατίσαι ἐπ' οἰκήματος, ἐντειλάμενον πάντας τε ὁμοίως προσδέκεσθαι, καὶ πρὶν συγγενέσθαι, ἀναγκάζειν λέγειν αὐτῇ ὅ τι δὴ ἐν τῷ βίῳ ἔργασται αὐτῷ σοφώτατον καὶ ἀνοσιώτατον· ὃς δ' ἂν ἀπηγήσηται τὰ περὶ τὸν φῶρα γεγενημένα, τοῦτον συλλαμβάνειν καὶ μὴ ἀπιέναι ἔξω.
>
> [The king] set up his own daughter in a bedroom,[40] instructing her to receive all men equally, but before sleeping with them to make each tell her what the cleverest and most impious things he had done in his life was. If anyone told her the thief's story, she was to grab hold of him and not let him [go] outside.

The wrongheadedness of this king in opening the door of his daughter's room to the thief mirrors Candaules' poor judgment in opening the door of his wife's room to Gyges, but what interests me more are the last three words of the passage: μὴ ἀπιέναι ἔξω ("not let him [go] outside"). Here, the word ἔξω (*out; outside*) takes us back to the story of Gyges, whose fate depended on finding a safe way out of doors (ἰόντα διὰ θυρέων) and situations (διαφυγεῖν; δυῶν ὁδῶν παρεουσέων), and who was described as ὑπεκδὺς ἐχώρεε ἔξω at 1.10.2. The prominent placement of ἔξω at the end of that sentence echoes its position at the end of the passage just quoted. The word that comes just before, ἀπιέναι, is also important because it is an infinitive of ἀφίημι (*to let go*) that shares the same form as the infinitive of ἄπειμι (*to go away, depart*). The success of Candaules' plan depends on the ability of the interloper to get outside; the success of Rhampsinitus' plan depends on the ability of the interloper not to do so. But in both, the issue is more specific than

40 This room (*oikêma*) is best understood as a brothel.

that. It is about reversing one's tracks and making one's way out around a door that should never have been opened to the outsider in the first place. In both, it is also about managing the business of undressing and carefully interlacing it with the business of exiting.

The coming and going of the interloper through doors that either cannot be discerned or, by contrast, that are opened too explicitly, forecasts a bedroom scene in which the balance of power is about to shift. In this bedroom, as with Candaules' room, we find the scene narrated twice – first through Rhampsinitus' instructions to his daughter, and shortly afterwards in Herodotus' own third person account. In the intervening period, the thief cuts the arm off a corpse and hides it under his cloak before going to visit the girl. When he has entered the room and truthfully answered the question posed to him, thereby suggesting that he has revealed his identity, the daughter of the king grabs him (2.121ε5):

> τὴν δέ, ὡς ἤκουσε, ἅπτεσθαι αὐτοῦ· τὸν δὲ φῶρα ἐν τῷ σκότεϊ προτεῖναι αὐτῇ τοῦ νεκροῦ τὴν χεῖρα· τὴν δὲ ἐπιλαβομένην ἔχειν, νομίζουσαν αὐτοῦ ἐκείνου τῆς χειρὸς ἀντέχεσθαι· τὸν δὲ φῶρα προέμενον αὐτῇ οἴχεσθαι διὰ θυρέων φεύγοντα.
>
> She, when she heard this, grabbed hold of him. But the thief in the dark stretched out to her the hand of the corpse. When she had seized it, thinking that it was his hand, she held onto it. But the thief releasing the hand to her, left, fleeing, through the doors.

The scene plays out as an inverse of what happened in Candaules' bedroom between Gyges and the queen. The door to this room stands wide open; the girl, instead of seeing the intruder clearly although her back is turned, sees him only imperfectly in the dark. While the queen waits, the princess grabs. And now, instead of one uniquely beautiful body being perfectly realized through the act of undressing, we find two bodies dressed up as a grotesque parody of one, as the girl uncovers a second arm beneath the first. As she takes hold of the thief in the imitation of an embrace, it is not the intimate act of the couple undressing that follows but rather, first, a sleight of hand played out under cover of the thief's clothes (an overdressing which is the reverse of an exposure), and then the man's exit from the room (οἴχεσθαι διὰ θυρέων φεύγοντα, "he left, fleeing through the doors"). The cleverness of the thief's exit looks back to the very first mention of passing through doors in the *Histories* (ἰόντα διὰ θυρέων, 1.9.2), initially stated by Candaules as if it were such a simple thing to do.

What is remarkable about the intruders of Rhampsinitus' rooms is that they are repeatedly able to escape identification. After the thief's exit, the king – *still* not knowing who the man who keeps entering his rooms is – declares an amnesty and offers him the hand of his daughter in marriage as a reward for revealing himself. Related to the problem of being exposed, then, is the problem of being recognized while in the room, a problem that the thief in Rhampsinitus' story manages to keep diverting but which other characters, such as Gyges, cannot.

Xerxes', Deioces', and Smerdis' Rooms

The thief's exceptional position in this regard contrasts with that of other men who enter as outsiders but at the devious behest of a king. These include Xerxes' uncle, Artabanus, who in book 7 reluctantly agrees to occupy the king's rooms by sitting on his throne, dressing up in his clothes, and sleeping in his bed. It is Xerxes' hope that the frightening dream that has been visiting him at night will be fooled into thinking that Artabanus is himself. The word used to describe the uncle's dressing in Xerxes' clothes, ἐνδύς (7.15, 17), emphasizes precisely the secret and illicit nature of his entrance into the king's chambers.[41] But, as Artabanus predicts, the dream is not fooled by this dressing up game, and the intruder finds himself not only exposed but also in a good deal of danger when the vision

41 Illicit because Artabanus, just like Gyges, balks at the idea of entering the personal rooms of the king, and has to be compelled to do so (7.16); secret because they are hoping to deceive the dream.

appears (7.17–18). Indeed, the fact that the dream almost burns Artabanus' eyes out with hot irons connects this bedroom scene with those others where the act of entering another man's room is explicitly connected to an illicit form of looking.[42]

No wonder, given the lessons that emerge from stories such as this, that the successful tyrant Deioces surrounds his palace with a series of circular walls and makes it impossible for nearly everyone else to enter (1.99). In order to get in and out of Deioces' room, a Mede would have to pass through seven doors instead of one. The protection that this provides is not just practical but also visual and, to a large extent, symbolic. By reinforcing the notion that the king's rooms, especially his bedroom, are private and secret spaces, Deioces underscores his own authority as a monarch.[43]

This is in some ways similar to what Smerdis tries to do when he fraudulently gains the Persian throne by pretending to be Cambyses' brother (3.68). He does not come out from the acropolis and calls none of the Persian nobleman into his sight. He summons his wives only at night, so that there is no opportunity for them to look at him even in his bedroom (3.68.4). While Deioces is working to maintain the idea of a secret in order to reinforce his rule, Smerdis is trying to hide the fact that he is pretending to be the rightful king (also named Smerdis). Otanes, suspecting as much, plots that his daughter, one of Smerdis' inherited royal wives, should enter the king's room in order to uncover his secret (3.69). Otanes successfully sidesteps the issue of entering the room himself (and thereby being exposed) by having his daughter legitimately do so instead. Specifically, he instructs her to feel beneath the

42 As understood by Gyges before he is compelled to enter Candaules' room (1.8.4: σκοπέειν τινὰ τὰ ἑωυτοῦ, "one should look to what is one's own").

43 Murnaghan forthcoming: "Deioces' strategy ... works through the manufacture of difference out of secrecy itself, through concealing the fact that there is nothing to hide." This also explains why the Persian Seven, in order to keep the balance of power in check, make a rule that any of them may enter the king's room at any time (unless he happens at that moment to be sleeping with a woman) (3.84.2).

hair of her husband when they are in bed together, to see if he has any ears.[44]

The danger and the secrecy involved in the plan to expose Smerdis, which might result in the daughter being exposed herself, is underscored by the use of ὑποδύνειν at 3.69.2: "My daughter, you are of noble birth and must engage in any danger that your father bids you to take on (ὑποδύνειν)." The verb prefigures not only the daughter's act of undressing in her husband's bedroom, but also his undressed state before her, which will enable her to feel for his ears under the cover of his hair (a second undressing and exposure). Elsewhere in Herodotus, ὑποδύ[ν]ω is applied to the putting on of tunics beneath cloaks (1.155) as well as to entering a tent by slipping in under it (4.75). In the context of this passage, both the clothing and entering/exiting valences apply, yet it is significant that because of Otanes' manipulation of who enters the room and under what circumstances, neither he nor his daughter is exposed (although she is put at risk of being so). There is no need for a "slipping out" once the bedroom has been entered. It is instead Smerdis who, once he has been so completely undressed and uncovered, will find it impossible to gain an exit from the room. For not long after this event he is chased into his bedroom and killed by Darius (3.78). As with Candaules, the site of Smerdis' demise is located within the room where he was involved in an earlier revelation and transgression.

It is worth noting that of the many meanings listed by the LSJ for ὑποδύ[ν]ω in addition to dressing and entering – including *slip from under* (I.4, s.v.), *put one's feet under a shoe* (II.1.b), *insinuate oneself into favor with* (II.1.d), *come from under* (II.2.1.c), and *slip or slink away* (II.2.5) – there is also the definition (LSJ s.v. II.1.c): "metaph., *put on* a character (because the actor's face was *put under* a mask) . . . " (Pl. *Grg.* 464c). The bedroom's overlaying of entering and exiting with the acts of dressing and undressing are perfectly fused in this meaning.

44 Otanes knows that the Persian magus Smerdis, whom he suspects is impersonating the king, has no ears. See further on this passage in Murnaghan forthcoming; Purves 2013: 34–36.

For the putting on and off of a mask speaks to the covering and uncovering of the self, but also of pretending to be someone else and occupying someone else's space. When the actor dons a mask he engages in a fiction that the space he enters is really his, and that particularly theatrical combination of costume with entrance and exit is one of the most important elements of ancient stage production. While the space of the performance gives us an open area in which to look, it is of course relevant, as noted earlier and as Padel has discussed, that the interior rooms behind the palace façade of the *skênê* remain closed to view. What Herodotus does in the *Histories* is to push the point further, by showing us how destabilizing it is to go inside, how hard it is to find a position from in there at which to look at the whole, and also – crucially – how impossible it is to see the bedroom at all without first making the disruptive step of entering it.

Gyges *was* able to watch unseen, of course, for as long as he remained in the subspace created by the retracted door. That sheltered but not completely closed-off space represents a small replica of the room itself, big enough for just one person and almost entirely secret. As we noted earlier, however, this space can only work as a hiding place within the room for as long as the door is left open. Any new entrance or exit has the potential to destroy it, exposing the person hidden within. Nevertheless, the temporary presence of this quiet corner of space thematizes the motifs of hiding, covering, and uncovering that are so prominent in Herodotus' rooms. In this sense, the place from which Gyges watches is an intermediary point between the room as a potentially open space (its door must open and shut, letting people in and out, no matter how dangerous that practice might ultimately be) and the room as a space that closes completely, like a chest that contains a secret or relates to special aspects of one's life. The chest is like the space behind the open door in that it acts as a miniature and intricately constructed double of the room, yet it remains closed and its interior unseen to everyone except the room's inhabitants.

Cypselus' Room

Both of these interior subspaces come into play in the story of Cypselus (5.92), who as a baby is sentenced to death by the Corinthians because of a double prophecy. The ten men who are secretly dispatched to carry out the job visit the baby and his mother inside her house. But with each man finding himself unable to kill the child as it is passed from hand to hand, they eventually step just outside the door to reproach one another. The mother then stands close behind the door (ἑστεῶσα πρὸς αὐτῇσι τῇσι θύρῃσι, 5.92δ.1.3) and overhears them speaking, thereby discovering the plot to kill her child. Again, therefore, we see that it is not just the threshold that matters, nor just the act of entering or exiting, but also the presence of a hidden observer or auditor, who – herself undetected – might uncover the secret plans or actions of others. It is not just that the door is important for its ability to open and close, in other words, but also for its ability to create a secret and temporary hiding space behind it, within which the dialectics of covering and uncovering, hiding and exposure, play back and forth.

With the realization that the men are about to come back into the room and kill the child, the mother hides him in a chest, thinking that it was the place "least likely to be thought of" (or "most unseen," ἀφραστότατον, 5.92δ.1.5) by the men, and when the murderers return and make a thorough search of the room, they do not find him. For "he was not visible" to them (5.92δ.2.2). Hidden within the chest, the child is covered (dressed, as it were) in so many layers as to escape detection by the men. The chest, like the coffin that enclosed Nitocris' corpse, thereby suggests a miniature, idealized version of a room that is bounded and secret, a place that remains unchanged and preserves its contents through time.[45]

[45] For the θηκή as a coffin that it is sacrilegious to open, cf. 3.16, 3.37, 4.71, as well as the Nitocris story discussed above (1.187). For the θηκή containing gold, see 3.130, 9.85. The θηκή is associated with a room (οἴκημα) in the descriptions of the burial vaults beneath the pyramids (e.g., 2.124.4.5–6). On the idea of the chest or box and its

Conclusion: Perdiccas and the Folds of Space

The last of Herodotus' anecdotes concerning rooms that I turn to in this chapter underscores the point that a room, unlike the closed space of a chest or box, always contains within itself the possibility of movement and change. In some ways the room, with its repeated mapping of the acts of dressing and undressing onto its own physical structure, is as mobile as the body itself and even portable enough to be carried in the folds of a cloak. We might think here not only of Alcmeon carrying out Croesus' gold but also of an ancestor of Alexander by the name of Perdiccas, a man of humble means who rose up to gain the Macedonian kingship through his understanding of the importance of space, as in the story that follows (8.137).

Perdiccas and his two older brothers were driven by necessity to leave their own country and work as outdoor laborers for the Macedonian king. When the wife of this king, however, told her husband that the loaf of bread she baked for Perdiccas each day rose to twice the size of the others, he suspected an omen and immediately ordered the three brothers off his land. When they reasonably refused to leave (ἐξιέναι) without first receiving their wages, the king – temporarily losing his wits – pointed to a circle of sunlight coming through a hole in his roof and told the brothers to take that as their pay instead (8.137.4):

> ἐνταῦθα ὁ βασιλεὺς τοῦ μισθοῦ πέρι ἀκούσας, ἦν γὰρ κατὰ τὴν καπνοδόκην ἐς τὸν οἶκον ἐσέχων ὁ ἥλιος, εἶπε θεοβλαβὴς γενόμενος "Μισθὸν δὲ ὑμῖν ἐγὼ ὑμέων ἄξιον τόνδε ἀποδίδωμι," δέξας τὸν ἥλιον.

> When the king heard this request for pay, since the sun was coming into the room through the hole in the roof for

representation of intimate space, see further Bachelard 1964: chapter 3 (74–89). These small boxes, chests, or caskets "are very evident witnesses of the *need for secrecy*, of an intuitive sense of hiding places" (1964: 81, emphasis original).

> the smoke, said, being misled by some god, "I give to you this as the thing worthy of you," pointing at the sun.

In response, Perdiccas has the presence of mind to accept what he is offered, by cutting out the space marked by the sunlight in the king's room and folding it into his clothing (8.137.5):

> 'Δεκόμεθα, ὦ βασιλεῦ, τὰ διδοῖς,' περιγράφει τῇ μαχαίρῃ ἐς τὸ ἔδαφος τοῦ οἴκου τὸν ἥλιον, περιγράψας δέ, ἐς τὸν κόλπον τρὶς ἀρυσάμενος τοῦ ἡλίου, ἀπαλλάσσετο αὐτός τε καὶ οἱ μετ' ἐκείνου.
>
> [Saying,] "We accept, king, the things that you give us," he circled with his knife around the sunshine on the floor of the house, and having cut it out, he drew the sunshine three times into the fold of his cloak. Then he left and the others went with him.

Perdiccas' performative act of dressing in the king's room and even clothing himself with the king's space is to be understood as a precursor to his subsequent entry to and takeover of the royal household.[46] Through the space of the room, Perdiccas is able to scoop up something of the king's power into his person, claiming the combined activities of "going out" and "dressing" as a tangible sign of his right to royal power. His actions, although they have much more sinister consequences, are similar to Alcmeon's clothing of himself in Croesus' gold. There, as I discussed above, the refined materiality of the gold merged with the qualities of space; here the space itself is imagined as perhaps a liquid substance, which can be drawn (ἀρύω) into Perdiccas' lap.[47]

The fold (κόλπος) into which Perdiccas thrice gathers the sunlit space of the king's room is also important for its ability to draw attention to space's material nature. For κόλπος is a word commonly used to describe a feature of the natural

46 This act is presented as the most significant detail in Herodotus' account of how Perdiccas became king.

47 LSJ *s.v.* ἀρύω, generally used of the drawing of liquids.

environment, such as a bay, gulf, or vale.[48] In the *Odyssey,* it frequently occurs in connection with the sea, and sometimes even in combination with entering and dressing, as in the following example, where Eidothea's covert act of dressing Menelaus and his men up in seal skins in order to get them access to the cave where her father sleeps pairs κόλπος (*fold of clothing; bay; lap*) with ὑποδύω (*go down under; put clothing on under*) (*Od.* 4.435–36):

> τόφρα δ' ἄρ' ἥ γ' ὑποδῦσα θαλάσσης εὐρέα κόλπον
> τέσσαρα φωκάων ἐκ πόντου δέρματ' ἔνεικε

> Meanwhile having gone down under the broad lap of the sea,
> she brought out of the water four seal skins.

The double clothing metaphor that lies beneath the surface of ὑποδύω and κόλπος, and which is further activated by the mention of seal skins for the men to wear in the following line, helps to reinforce the point that the sea is a secret and private family space for Eidothea, which she is here breaching by her actions (facilitating the ambush of her father). The enclosure created by the fold is like the enclosure of a room, but it is also associated further with the idea of intimate space through its connection with bosom and lap.[49]

Thinking about space having materiality in this way is, therefore, a helpful means of interpreting the movement between the states of being dressed and undressed, covered and exposed, within the space of the room. Candaules' wife's initial act of undressing was, as we first discussed, mirrored in the structure of the room in which she slept, both through the careful attention Candaules draws to the furniture around her and through the manner of Gyges' exit. The bedroom's role as a place that is as private, and perhaps as vulnerable, as the naked human body sets it up as a unique kind of viewing space. To

48 LSJ *s.v.* III. This compares to the English use of the word, as in the "folds" of a mountain.

49 LSJ *s.v.* I.1, 2. Eidothea's betrayal consists of getting Menelaus, in disguise, into what is more or less Proteus' bedroom.

look at a naked body and to look at an area of space are manifestly not the same thing, yet Herodotus combines the two within the walls of the bedroom. By inviting characters within his work to step over the threshold of the bedroom when they should not, he plays with the established Greek idea that what is inside must also be "covered" and hidden from sight.[50]

The juxtaposition between entering or exiting the personal space of the room and (un)dressing shows up enough times in the *Histories* to suggest that these rooms belong to kings who all try to maintain an uneasy – and ultimately impossible – alliance between intimacy and power. The extensive geographical reach of the *Histories* and the importance of its central characters in large-scale historical events makes the small space of the room all the more significant. The tyrant in Herodotus may be, like Xerxes in this chapter's opening example, the master of all he surveys, but this stance can pose serious problems when he trains his look on his own interior spaces. From the point of view of Herodotus as a historian, this suggests that different spaces are describable in different ways. On the one hand, the sources of the Nile might be impossible to discover or the far reaches of Scythia impossible to recount, but what we also discover here is that those rooms which are small enough to be fully described pose their own ambivalent relationship to the opening up and disclosing of space to view. When Herodotus takes us into these bedrooms he hints at a special kind of looking into worlds that are emphatically not his readers' own, but as he does so he also tacitly acknowledges that these rooms only appear in his stories because their owners could no longer hold onto them as fixed and private spaces.[51]

Through these narratives about bedrooms, Herodotus reverses the stereotype that situates a transgressive (Greek) woman standing at the door and looking or moving *out*; the women in these stories instead act only reciprocally, when

50 Cairns 2002; Llewellyn-Jones 2003.

51 Cf. Christ 1994, on parallels between Herodotus and his barbarian "inquiring" kings.

their space is violated first.[52] It is not just, then, that barbarian kings take pleasure in viewing,[53] it is – beyond that – that their rooms are situated on a fault line of power, secrecy, and desire that the gaze of an outsider specifically triggers. And, in general, these kings cannot resist allowing outsiders to look within. By taking us into the king's room and opening it up to view, Herodotus troubles its status as a space that is supposed to be closed to the public eye. In contrast to a landscape, as far as the special interior space of these rooms are concerned, there is no ideal or stable position to view it from the outside without coming in.[54]

Bibliography

Bachelard, G. 1964. *The Poetics of Space*. Boston.

Benardete, S. 1999. *Herodotean Inquiries*. South Bend, IN.

Bewes, T. 2005. "What is a Literary Landscape? Immanence and the Ethics of Form." *Differences* 16.1: 63–102.

Cairns, D. L. 1996. "'Off with Her ΑΙΔΩΣ': Herodotus 1.8.3–4." *CQ* 46.1: 78–83.

2002. "The Meaning of the Veil in Ancient Greek Culture." In *Women's Dress in the Ancient Greek World*, ed. L. Llewellyn-Jones. Swansea. 73–93.

2012. "Vêtu d'impudeur et enveloppé de chagrin. Le rôle des métaphores de 'l'habillement' dans les concepts d'émotion en Grèce ancienne." In *Les vêtements antiques: s'habiller, se déshabiller dans les mondes anciens*, ed. F. Gherchanoc and V. Huet. Paris. 149–62.

Casey, E. S. 1997. *The Fate of Place*. Berkeley.

Christ, M. R. 1994. "Herodotean Kings and Historical Inquiry." *CA* 13.2: 167–202.

Cosgrove, D. E. 1985. "Prospect, Perspective and the Evolution of the Landscape Idea." *Trans. Inst. Br. Geogr.* 10: 45–62.

Evans, J. A. S. 1955. "Herodotus and the Gyges Drama." *Athenaeum* 33: 333–36.

Ferrari, G. 1990. "Figures of Speech: The Picture of *Aidos*." *Métis* 5.1–2, 185–204.

52 On the question of the "female intruder" (the woman who intrudes on the public sphere) in Greek drama, see Foley 1982.

53 Konstan 1987.

54 I wish to thank Douglas Cairns, Sheila Murnaghan, and Mario Telò, as well as the editors and readers of this volume, for their helpful comments on this paper.

Flory, S. 1987. *The Archaic Smile of Herodotus*. Detroit.

Foley, H. P. 1982. "The 'Female Intruder' Reconsidered: Women in Aristophanes' *Lysistrata* and *Ecclesiazusae*," *CP* 77.1: 1–21.

Konstan, D. 1987. "Persians, Greeks and Empire." *Arethusa* 20.1/2: 59–73.

Kurke, L. 1999. *Coins, Bodies, Games, and Gold: The Politics of Meaning in Archaic Greece*. Princeton.

Lefebvre, H. (1974) 1991. *The Production of Space*. Oxford.

Llewellyn-Jones, L. 2003. *Aphrodite's Tortoise: The Veiled Woman of Ancient Greece*. Swansea.

Lobel, E. 1950. "A Greek Historical Drama." *Proceedings of the British Academy* 35: 1–12.

Mitchell, W. J. T. (ed.). 2002. *Landscape and Power*. Chicago.

Munson, R. V. 1993. "Herodotus' Use of Prospective Sentences and the Story of Rhampsinitus and the Thief in the *Histories*." *AJP* 114.1: 27–44.

Murnaghan, S. H. Forthcoming. "Naming Names, Telling Tales: Sexual Secrets and Greek Narrative." In *Sex in Antiquity: Exploring Gender and Sexuality in the Ancient World*, ed. M. Masterson, N. Sorkin Rabinowitz, and J. Robson. London.

Padel, R. 1990. "Making Space Speak." In *Nothing To Do With Dionysos? Athenian Drama in its Social Context*, ed. J. J. Winkler and F. I. Zeitlin. Princeton. 336–65.

Page, D. 1951. *A New Chapter in the History of Greek Tragedy*. Cambridge.

Purves, A. C. 2010. *Space and Time in Ancient Greek Narrative*. New York and Cambridge.

2013. "Haptic Herodotus." In *Synaesthesia and the Ancient Senses*, ed. S. Butler and A. Purves. Durham, U.K. 27–41.

2014. "Sleeping Outside in Homer's *Odyssey* and W. G. Sebald's *The Emigrants*." In *Odyssean Identities in Modern Cultures: The Journey Home*, ed. H. Gardner and S. Murnaghan. Columbus, OH 213–39.

Raubitschek, A. E. 1955. "Gyges in Herodotus." *The Classical Weekly* 48.4: 48–50.

Romm, J. S. 1992. *The Edges of the Earth in Ancient Thought: Geography, Exploration, and Fiction*. Princeton.

Rood, T. 2006. "Herodotus and Foreign Lands." In *The Cambridge Companion to Herodotus*, ed. C. Dewald and J. Marincola. Cambridge. 290–305.

Soja, E. W. 1996. *Thirdspace: Journeys to Los Angeles and Other Real-and-Imagined Places*. Cambridge, MA.

Taplin, O. 1977. *The Stagecraft of Aeschylus: The Dramatic Use of Exits and Entrances in Greek Tragedy*. Oxford.

Thalmann, W. G. 2011. *Apollonius of Rhodes and the Spaces of Hellenism*. New York and Oxford.

Travis, R. 2000. "The Spectation of Gyges in P. Oxy. 2382 and Herodotus Book 1." *CA* 19.2: 330–59.

Tuan, Y.-F. 1974. *Topophilia: A Study of Environmental Perception, Attitudes, and Values*. Englewood Cliffs, NJ.

1977. *Space and Place: The Perspective of Experience*. London.

2004. "Sense of Place: Its Relationship to Self and Time." In *Reanimating Places: A Geography of Rhythms*, ed. T. Mels. London. 46–55.

Vernant, J.-P. 1983. *Myth and Thought Among the Greeks*. London.

West, S. 2007. "Rhampsinitos and the Clever Thief (Herodotus 2.121)." In *A Companion to Greek and Roman Historiography Volume II*, ed. J. Marincola. Cambridge, MA. 322–27.

Wylie, J. 2006. "Depths and Folds: On Landscape and the Gazing Subject." *Environment and Planning D: Society and Space* 24: 519–35.

4

SHIPS, WALLS, MEN: CLASSICAL ATHENS AND THE POETICS OF INFRASTRUCTURE

Carol Dougherty

> Certainly, when we build we are telling stories about the world, sculpting the cultural landscape even as we remold the physical one. But if buildings tell stories, it is also true that stories make buildings. When office, stores, and homes are suddenly and unexpectedly annihilated, it is necessary not only to manufacture new material structures but also to repair torn cultural fabrics and damaged psyches. With this in mind, I propose to explore the relationship between the rebuilding of cities with mortar and bricks and the rebuilding of cultural environments with words and images in the aftermath of great urban disasters – a double process neatly caught in the twin meanings of the word *reconstruction* as "remaking" and as "retelling."
>
> K. Rozario, "Making Progress. Disaster Narratives and the Art of Optimism in Modern America"

In 480 BCE, deserted by their allies, the Athenians chose to abandon their homeland rather than stay and face a threatening Persian army on land. After sending their women and children abroad, they put their faith in their fleet and famously defeated Xerxes and the Persians in a sea battle at Salamis.[1] Victorious,

[1] Cf. Strauss 2004 for a dramatic account of these events.

they returned, however, to a city in ruins, utterly destroyed by the Persians who had occupied Athens in their absence. Thucydides describes the sight facing them, "of the perimeter wall only small portions were still standing, and most of the houses had fallen, only a few remained in which the Persian leaders had themselves taken quarters."[2] Herodotus tells us that "he [Xerxes] burned Athens, and whatever walls or houses or temples were still upright, he destroyed and demolished."[3] And yet in the wake of this devastating destruction, the Athenians literally rebuilt their city. Not only did they replace their houses, they immediately and hastily repaired the damaged circuit of walls around the old city. Subsequently, they connected this fortified city to a new and improved port in the Piraeus via the Long Walls, a decision that radically transformed the Athenian civic experience on many levels from the material to the imaginary. The deep harbors and elaborate ship sheds at Piraeus made possible the extensive naval force with which Athens ruled the Mediterranean as a successful imperial power for the next half century, and Athens' city walls, both the Long Walls connecting city and port and the newly repaired walls encircling the old city, came to define Athens as an increasingly urban city in the fifty years since the Persian wars.[4] In other words, to borrow terminology used by cultural geographers such as David Harvey and Tim Cresswell, the Athenians transformed the devastated, uninhabited *space* left behind by the defeated Persians into a newly configured civic *place* – a prosperous naval empire.[5] In particular, the ships

2 Thuc. 1.89.3.

3 Hdt. 9.13. For the archeological evidence, see Thompson 1981; Shear 1993.

4 On the rebuilding of the fortification walls, cf. Thuc. 1.89.3–93.2. Thucydides mentions the decision to follow Themistocles' advice to fortify the new harbor at Piraeus at 1.93. In 458/7 the Athenians decided to link the city to the port and embarked upon a project to build the Long Walls; cf. Conwell 2008 for history of the Long Walls and for ancient sources.

5 As noted in the introduction to this volume, recent scholarship in cultural geography, philosophy, and literary criticism explores issues of spatial contingencies by articulating the difference between space

and walls with which the Athenians restored their physical city also served as useful tools for reconstructing Athens' cultural environment, and this essay explores the different ways in which several key texts from the last third of the fifth century participate in the remaking of the classical Athenian city as a literary *topos*, not just a physical place.[6] As cities rise and fall, literary texts are especially adept at making and remaking the civic imaginary, at repairing "torn cultural fabrics and damaged psyches."[7] The following discussion reads passages from Herodotus' *Histories*, Sophocles' *Oedipus the King*, and Thucydides' *History of the Peloponnesian War* to chart the significant interrelationships between ships, walls, and the men who occupy them within what I have termed a "poetics of infrastructure."[8] All three texts date to the last third of the

and place. For most, *space* and *place* occupy opposing ends of the spatial continuum: *space* is the more abstract concept while *place* is associated with experience, with people and the human capacity to produce and consume meaning. Cf. Harvey 1996; Cresswell 2004: 15–51 offers a useful overview of recent scholarship on *place* and *space*. De Certeau 1994: 117 uses these terms but reverses them in his article on "Spatial Stories" to argue that "space is a practiced place."

6 Cf. de Certeau 1994: 115–30, who shows the ways that literary texts contribute to the construction of place; he argues that narrative is a spatial practice, one which is key to the transformation of spaces into place or places into spaces: "[Stories] also organize the play of changing relationships between places and spaces" (118).

7 As expressed by Rozario in the epigraph to this essay. See also Porter 2011: 33 whose discussion of the "literariness" of the Achaean wall in the *Iliad* focuses on the wall as a sublime object, one that allows us to engage with the mutual implication of its poetic and physical construction: "The Achaean wall is, in other words, a deeply strange object that defeats our expectations. Yet it is not a symbol of the impermanence of things pure and simple, let alone of writing or song. Rather it is a *sublime* object that is endowed with an indelible *kleos*, a lasting fame that lives on, not even if the wall vanishes, but precisely *because* the wall vanishes – for as long as the memory that recalls it persists, wanted or not."

8 I have borrowed the term "poetics of infrastructure" from an essay by Yeager 2007: 21, in which she argues that just as Aristotle's *Poetics* articulates a systematic account of the structure and flow of Greek tragedy, so, too should there be a "poetics of infrastructure or metropoetics," that would, as she explains it, "describe the flow of

fifth century, a time when Athenians were once again threatened militarily and once again reflecting upon how to defend and define their city. What will emerge from these readings, then, is not a clear, coherent picture of the role that ships and walls played in defining the Athenian civic experience, but rather a repeated return to these figures, combined in different ways with different associations, in an ongoing and evolving attempt to accommodate the radical changes in the shape of the city within the civic imaginary. In particular, what will become clear is the inextricability of ships and walls as competing and complementary figures for the city at this time. They form a dyad, functioning as a kind of polarizing doublet that enables the "reconstruction" of Athens as both a city of walls and a newly configured naval empire in the fifty years immediately following the Persian destruction of the city.

HERODOTUS' *HISTORIES*: "A WALL OF WOOD"

Let's start with Herodotus' *Histories*, a text from the 420s that tells the story of that defining moment more than half a century earlier, when the Athenians had to decide how best to defend themselves against the impending Persian invasion.[9] Toward the end of his *Histories*, Herodotus explains that the Athenians consulted the oracle at Delphi (twice) about defense strategies to be used against the Persians, and his text provides our best source both for the oracles they received as well as for the interpretative debate that the famous "wooden wall" oracle generated back in Athens, a debate that played a formative role in the Athenian cultural memory of the Persian

literature as it pours into and out of the life of cities." My own use of the term will focus as much on the word "infrastructure" as on "poetics," arguing for the specific ways in which radical changes in Athens' infrastructure (ships and walls) forms the essential scaffolding, as it were, for the cultural "reconstruction" of Athens in the texts under discussion.

9 For the date of Herodotus' text, I agree with Thomas 2006: 61 that he probably was working on the *Histories* over a period of time, with the final publication as late as the 420s. See also Fornara 1971.

wars throughout the fifth century.[10] Here is the first oracle that Herodotus records:

> ὦ μέλεοι, τί κάθησθε; λιπὼν φεῦγ᾽ ἔσχατα γαίης
> δώματα καὶ πόλιος τροχοειδέος ἄκρα κάρηνα.
> οὔτε γὰρ ἡ κεφαλὴ μένει ἔμπεδον οὔτε τὸ σῶμα,
> οὔτε πόδες νέατοι οὔτ᾽ ὦν χέρες, οὔτε τι μέσσης
> λείπεται, ἀλλ᾽ ἄζηλα πέλει· κατὰ γάρ μιν ἐρείπει
> πῦρ τε καὶ ὀξὺς Ἄρης, Συριηγενὲς ἅρμα διώκων.
> πολλὰ δὲ κἆλλ᾽ ἀπολεῖ πυργώματα, κοὐ τὸ σὸν οἶον·
> πολλοὺς δ᾽ ἀθανάτων νηοὺς μαλερῷ πυρὶ δώσει,
> οἵ που νῦν ἱδρῶτι ῥεούμενοι ἑστήκασι,
> δείματι παλλόμενοι, κατὰ δ᾽ ἀκροτάτοις ὀρόφοισιν
> αἷμα μέλαν κέχυται, προιδὸν κακότητος ἀνάγκας.
> ἀλλ᾽ ἴτον ἐξ ἀδύτοιο, κακοῖς δ᾽ ἐπικίδνατε θυμόν.

> Wretched ones, why sit you here? Flee and be gone to remotest Ends of earth, leaving your homes, high places in circular city; for neither the head abides sound, no more than the feet or the body; fire pulls all down, and sharp Ares, driving his Syrian-bred horses. Many a fortress besides, and not yours alone shall he ruin. Many the temples of God to devouring flames he shall give them. There they stand now, the sweat of terror streaming down from them. They shake with fear; from the rooftops black blood in deluging torrents. They have seen the forthcoming destruction, and evil sheerly constraining. Get you gone out of the shrine! Blanket your soul with your sorrows. (Hdt. 7.140; Grene trans.)

10 Herodotus' account of the events leading up to (and including) the unexpected Athenian victory over a superior Persian force at Salamis has been thoroughly discussed, with particular attention to questions of historicity. See, for example, Evans 1982 (includes a review of the arguments for dating the oracles themselves); Georges 1986; Holladay 1987; Robertson 1987. The discussion that follows will focus not on the historical accuracy of Herodotus' account, but rather on the language and structure of his representation of this particular Delphic consultation. See Euben 1986: 360 for a discussion of how the Salamis narrative "set the terms in which Athenians defined themselves as a people."

The oracle begins with a clear admonition to flee; it urges the Athenians to leave behind their houses (δώματα) and the city's fortification (πόλιος τροχοειδέος ἄκρα κάρηνα) because the city will be completely destroyed by fire and war, and it offers a gruesome and surreal account of a city under siege.

As Lisa Maurizio has noted, the language of this oracle, with its ominous imagery of decapitation and dismemberment, offers an embodied view of the city, one that physically identifies the city with its land, its territory, the altars and sanctuaries of its gods.[11] The metaphorical use of the word head (κάρηνα) to represent the city's tallest structures then leads to an elaborate catalogue that represents the impending destruction of the city in human terms, literally in terms of body parts. Nothing will remain of the embodied city – not the head, the body, the feet, the hands, nor anything left in the middle. For fire and war will destroy the city's defenses and its temples. Moreover the oracle extends the animating metaphor by giving the city human emotions and experiences: its temples drip with sweat, shake with fear, and shed blood just as would their human occupants in a state of siege.

The imagery of this first oracle, with its graphic description of the destruction of the city as body, depends upon the terms of a well-established analogy between a city's men and its physical fortifications. The archaic poet, Alcaeus, for example, equates the fierceness of a city's men with its physical defenses:

> ἄνδρες γὰρ πόλιος πύργος ἀρεύιοι
>
> warlike men are a city's tower (fr. 112.10 LP)

As Aelius Aristides glosses the passage, "cities are not stones or timbers or the craft of builders; but wherever there are men who know how to defend themselves, there are walls and cities."[12] The word order of the Alcaeus passage reinforces the

11 Cf. Maurizio 1997: 330–31.

12 Ael. Arist. *Or.* 46.207 = Alc. Fr. 426 LP. Cf. Morris 1987: 4–5, who cites this passage (together with Thuc. 7.77.7) as evidence for the

poet's point: the two words describing the warlike men (ἄνδρες; ἀρεύιοι) encircle the words for city (πόλιος) and tower (πύργος) just as city walls literally surround the city. Here, the word for city is in the exact middle of the line, protected on both sides by brave men and towers, a subjective genitive belonging equally to both. Although as Aelius Aristides explains the passage, the fragment privileges the defensive powers of brave men over mere physical walls as truly constitutive of the city, the lack of verb or any syntactic subordination (a predicate nominative construction) in the Alcaeus fragment leaves us in the end with a tripartite set of associations by which the city belongs equally to or is equally made up of its men and its defensive structures.

Similarly, Aeschylus in the *Persians*, a play performed a mere ten years after the Persians sacked Athens, reworks Alcaeus' formulation with specific respect to Athens to describe that city's invincibility in the face of the Persian invasion. In the *Persians*, the Persian queen, Atossa, asks a messenger to tell her the news from Athens. Upon hearing the devastating news of the Persian defeat, she is incredulous and asks whether Athens remains standing:

> ΑΤ. ἔτ' Ἀθηνῶν ἔστ' ἀπόρθητος πόλις;
> ΑΓ. ἀνδρῶν γὰρ ὄντων ἕρκος ἐστὶν ἀσφαλές.
>
> Atossa: "Is the city of Athens still unsacked?"
> Messenger: Yes, for while her men still live, her defenses are secure. (Aesch. *Per*. 347–49)[13]

long-standing practice of associating men with the polis: "Across these centuries, the *koinonia* was the heart of the polis." Part of what I argue here is that in Athens in the last half of the fifth century, the third term (walls) of what Alcaeus articulates as a three part analogy (city = walls = men) is challenged by Athens' emerging identity as a city of ships.

13 Harrison 2000: 93 argues that this passage from the *Persians* (together with a remark attributed by Herodotus to Themistocles about being a man without a city which I will discuss below) depend on the "Wooden Wall" oracle. It is interesting to note that ἀπόρθητος is the word used to describe the impregnability of the wooden wall in the second oracle cited by Herodotus, as discussed below.

Here Aeschylus leads with men (ἀνδρῶν γὰρ ὄντων), the syntactic independence of the genitive absolute underscoring the role of the men as sufficient symbols of the city's enduring strength. Again, as in the fragment from Alcaeus, the logic of Aeschylus' image first equates the city walls with the city and then the city's men with its walls. As long as Athens' men remain strong, the city survives.[14]

Returning to Herodotus' first oracle, we see that the language used to describe the city under siege belongs to this familiar set of imagery defining a city by its physical infrastructure and equating it with the strength and bravery of its men. The oracle makes the analogy all the more striking, however, by inverting the tenor and vehicle of the metaphor. Whereas both Alcaeus and Aeschylus assert that men are the city's walls, this first oracle describes the city's physical defenses (πόλιος τροχοειδέος ἄκρα κάρηνα) as if they were men with a head, body, feet, and hands (ἡ κεφαλή; τὸ σῶμα; πόδες; χέρες). And by reversing the terms of the analogy, the oracle highlights the vulnerability of the city to the devastating external forces of fire and war rather than celebrating its invincibility, as do the passages from Alcaeus and Aeschylus. This inversion with its emphasis on the city at risk leads naturally into the second oracle, which, as we will see, introduces a very different image for the city.

The terrifying images of this first oracle led the Athenians to return as suppliants and beg the prophetess for another oracle, which Herodotus records as follows:

οὐ δύναται Παλλὰς Δι' Ὀλύμπιον ἐξιλάσασθαι,
λισσομένη πολλοῖσι λόγοις καὶ μήτιδι πυκνῇ.
σοὶ δὲ τόδ' αὖτις ἔπος ἐρέω, ἀδάμαντι πελάσσας·
τῶν ἄλλων γὰρ ἁλισκομένων ὅσα Κέκροπος οὖρος
ἐντὸς ἔχει κευθμών τε Κιθαιρῶνος ζαθέιοιο,

14 Even if you take the genitive absolute to represent a conditional clause, the sense remains the same: if the men remain strong, the city survives. By contrast with Susae which, as Rosenbloom 1993: 191 observes, is the very opposite: a city of walls without men.

τεῖχος Τριτογενεῖ ξύλινον διδοῖ εὐρύοπα Ζεὺς
μοῦνον ἀπόρθητον τελέθειν, τὸ σὲ τέκνα τ' ὀνήσει.
μηδὲ σύ γ' ἱπποσύνην τε μένειν καὶ πεζὸν ἰόντα
πολλὸν ἀπ' ἠπείρου στρατὸν ἥσυχος, ἀλλ' ὑποχωρεῖν
νῶτον ἐπιστρέψας· ἔτι τοί ποτε κἀντίος ἔσσῃ.
ὦ θείη Σαλαμίς, ἀπολεῖς δὲ σὺ τέκνα γυναικῶν
ἤ που σκιδναμένης Δημήτερος ἢ συνιούσης.

No: Athena cannot appease great Zeus of Olympus
With many eloquent words and all her cunning counsel.
To you I declare again this word, and make it as iron:
All shall be taken by foemen, whatever within his border
Cecrops contains, and whatever the glades of sacred Cithaeron.
Yet to Tritogeneia shall Zeus, loud-voiced, give a present,
A wall of wood, which alone shall abide unsacked by the foemen; Well shall it serve yourselves and your children in days that shall be. Do not abide the charge of horse and foot that come on you, A mighty host from the landward side, but withdraw before it. Turn your back in retreat; on another day, you shall face them, Salamis, isle divine, you shall slay many children of women, Either when seed is sown or again when the harvest is gathered. (Hdt. 7.141; Grene trans.)

While this second oracle does not predict a radically different end result for the Athenians, it does offer a significantly different way of thinking about the city – one that is embodied in the small glimmer of hope that Zeus gives to Athena: the wooden wall that alone will remain unsacked.

At the beginning of this oracle, the captured city is first expressed physically as the territory within Athens' borders rather than as a dismembered body: "whatever within his border/ Cecrops contains, and whatever the glades of sacred Cithaeron." In this formulation, the city is defined by the land itself, not public buildings or temples. The invocation of both Athena Tritogeneia and Cecrops appeals to Athens' civic traditions of autochthony, being born from the earth, through which Athenians celebrate their privileged status as indigenous inhabitants of their land, boasting that they are neither immigrants from another city nor exiles to another land.

Mention of the wooden wall falls almost exactly in the middle of the oracle, and the disputed phrase marks the key point at which the enigmatic language of the second oracle pivots from one way of thinking about the city to another. Themistocles' ultimately successful interpretation of the "wooden wall" as ships capable of fighting at sea is actually foreshadowed within the oracle itself by the subsequent dismissal of the other possible modes of military defense. The Athenians should not wait to meet the Persians with the cavalry (ἱπποσύνην) nor with their landed infantry (πεζὸν ἰόντα / πολλὸν ἀπ' ἠπείρου στρατὸν).[15] Instead, reading the walls as ships resolves what initially looked like a typically Delphic oxymoron: retreat and turn your backs (ἀλλ' ὑποχωρεῖν / νῶτον ἐπιστρέψας) and yet still you will be face to face (ἔτι τοί ποτε κἀντίος ἔσσῃ). In other words, the Athenians turn their backs on the embodied notion of their city; they evacuate their city and then defend it successfully against the Persians at sea.[16] The Athenians do, of

15 Cf. Mardonius' speech encouraging Xerxes to continue to fight against the Athenians (and not run away in defeat), in which he uses the same terms to argue against fighting by sea and in favor of a contest with cavalry: "For the contest that carries all and everything for us is not one of wood but of men and of horses" (οὐ γὰρ ξύλων ἀγὼν ὁ τὸ πᾶν φέρων ἐστὶ ἡμῖν, ἀλλ' ἀνδρῶν τε καὶ ἵππων), Hdt. 8.100.2.

16 In this respect, the oracle gives the Athenians language for rewriting what looks like a retreat or surrender – abandoning one's city – as an aggressive military gesture – fighting face to face at with their ships. In addition, the second oracle addresses the inherent contradiction in the notion that the Athenians can save their city by abandoning it. For clearly another possibility is the more straightforward option of abandoning Athens permanently and resettling elsewhere. As Herodotus notes at 7.143, some of the oracle interpreters were, in fact, advocating complete withdrawal: "[they] would not have them prepare for a sea fight nor even, to tell the whole story, to put up any resistance at all, but they should leave Attica and settle in some other country."

course, evacuate their city (8.41), and the Persians sack it and destroy its physical structures – its buildings and walls – thoroughly (8.53; 9.13). But, as Herodotus subsequently reports, the city that the Persians capture is a deserted one (καὶ αἱρέουσι ἔρημον τὸ ἄστυ, 8.51; cf. 9.3), and its capture does nothing to bring about the defeat of the Athenians. For, in this reading, Athens is no longer defined by its city walls, but as something else much more mobile – its ships.

Later in book 8 of his *Histories*, Herodotus makes explicit the new civic model that this oracle suggests – a city of ships. Prior to the battle at Salamis, the Corinthian Adeimantus challenged Themistocles' right to leadership of the Greek coalition on the grounds that Themistocles was a cityless man (ἀπόλι ἀνδρί) because Athens had been captured and was currently occupied by the Persians. But Themistocles replied that the Athenians had a city and land superior (πόλις καὶ γῆ μέζων) to that of the Corinthians, "as long as they had two hundred fully manned ships" (ἔστ' ἂν διηκόσιαι νέες σφι ἔωσι πεπληρωμέναι, 8.61.2).[17]

Associating the city with a ship, a rhetorical figure traditionally known as the "ship of state," is of course also familiar from archaic Greek (and other) poetry.[18] Indeed, it is once again Alcaeus to whom we turn for an early example of this trope. He represents the tumultuous political situation of archaic Mytilene in terms of a storm-tossed ship amidst threatening waves:

> ἀσυν⟨ν⟩έτημμι τῶν ἀνέμων στάσιν·
> τὸ μὲν γὰρ ἔνθεν κῦμα κυλίνδεται,
> τὸ δ' ἔνθεν, ἄμμες δ' ὂν τὸ μέσσον
> νᾶι φορήμ⟨μ⟩εθα σὺν μελαίνᾳ
>
> χείμωνι μόχθεντες μεγάλῳ μάλα·

17 Cf. Aristoph. *Birds* 108 where Pisthetaerus also characterizes Athens as a city of ships; when asked where he is from, he replies, "where the beautiful triremes are (ὅθεν αἱ τριήρεις αἱ καλαί).

18 Cf. Taylor 2010: 142–43; Pelling 2000: 16; Thompson 2001: 167–72 for additional examples of the ship of state image, both generally and as it has been used specifically for classical Athens.

I do not understand the conflict of the winds
For one wave rolls in from here
And another from there, and we in the middle
Are carried along with our black ship

Suffering much in the big storm . . . (fr. 208.1–5 L-P)[19]

The use of stasis here to describe the unruly storm winds calls to mind its familiar metaphorical use to refer political crosswinds and sets the stage for reading the subsequent account of a ship tossed upon savage seas to represent the tumultuous political situation in similar terms. As the ship of state rolls from one political extreme to another, the poet and his companions hunker down in the middle (τὸ μέσον), hoping to endure the storm. The designation of the middle of the ship as a political position appears in a passage from an early Athenian context as well. In his account of Solon's ability to navigate a smooth course between the competing political demands of the wealthy property owners and the poor in early sixth century Athens, Plutarch records the following Delphic oracle, allegedly delivered to Solon:

ἧσο μέσην κατὰ νῆα κυβερνητήριον ἔργον
εὐθύνων· πολλοί τοι Ἀθηναίων ἐπίκουροι.

Sit in the middle of the ship, steering a straight course;
there are many Athenian helpers.[20]

Here, as in the Alcaeus passage, the middle of the ship comes to represent a centrist political agenda, and both the Alcaeus and the Solon passage emphasize the other occupants of the ship; in Alcaeus, it is a plural subject that is huddled into the bowels of the ship, waiting out the storm, and Solon's oracle reminds

19 We find a similar set of images in another fragment of Alcaeus (6 L-P), in which the speaker laments the faltering ship of state and urges his companions to strengthen the sides of the ship and run toward safe harbor (ἐς δ' ἔχυρον λίμενα δρό[μωμεν; 6.8). Cf. Page 1955: 181–82 on the ship of state image in Alcaeus.

20 Plut. *Solon* 14.6. This oracle is included in the collections of both Parke and Wormell 1956 (#15) and Fontenrose 1978 (Q67).

him encouragingly that he has many Athenian allies aboard his ship of state.[21] In this passage, the ship imagery both signals a kind of buoyancy in times of stress and underscores a collective, corporate sense of citizenship, shared by all, with Solon at the helm. It is certainly this traditional literary trope, then, that Herodotus' Themistocles draws upon when he boasts that as long as Athens has its ships fully manned, the city lives on. In late fifth-century Athens, however, as Herodotus' audience knows all too well, the ship of state trope gains even more potent metaphorical force from Athens' own particular experience at Salamis and all that transpired as a result of decisions made to develop Athens' navy at that time.

In other words, the two interpretations of "wooden wall," as outlined in Herodotus' narrative of the subsequent debate back in Athens, represent two different ways of thinking about the city, both of which focus on key elements of civic infrastructure. The reading of the wooden wall as the thorn hedge that once fenced in the Acropolis, put forth by the elder Athenians, depends upon the familiar notion (articulated in Alcaeus, Aeschylus, and the first oracle) that the city is defined by its built environment, especially its walls, while Themistocles' interpretation of the wooden wall metaphorically, as ships, hinges upon a very different notion of the definitive infrastructural elements at the core of the city (suggested by Alcaeus, Solon, and the second oracle). While both metaphors highlight the significance of men for the survival of the city, the oracle's successful interpretation hinges

21 Cf. Silk 1974: 122–24 for a discussion of the "ship of state" figure in both the Solon and Alcaeus passages we are discussing. In light of the wooden wall oracle, it is interesting that this association of the ship of state image with Solon, a founding figure of the Athenian state, has also been appropriated within a Delphic narrative. The ship of state image continues to be a popular one in Athenian literature. See, for example, Aristophanes *Wasps* 29, where Sosias says his dream is about the ship of state (περὶ τῆς πόλεως γάρ ἐστι τοῦ σκάφους ὅλου), and the scholiast on the passage remarks: "poets are always comparing cities to ships." Cf. Koster 1975, cited in Taylor 2010: 142n20.

upon which physical attributes will first defend and then define the city. Will walls or ships serve them best? The plot of Herodotus' narrative thus uses the ambiguous oracle with its timely and successful interpretation to articulate a reformulation of Athens' civic identity in conjunction with these changes in the city's built environment. It helps the Athenians articulate a shift away from conceptualizing their city in terms of its land and physical structures to be defended. In its place, Herodotus' narrative offers the image of a ship whose mobility and military capabilities can save the city in different ways, a metaphor that certainly reflects the real experience of what has happened to the city of Athens as it developed its naval empire in the fifty years since the Persian Wars.

And yet, the narrative enacts this conceptual shift all the while refusing to erase the memory of older ways of thinking about the city. Herodotus uses the Delphic framework to play with time, rearranging the recent history of the Athenian city into a synchronic kaleidoscope of competing models of civic infrastructure that challenge the primacy and inevitability of any one of them.[22] The city that the Athenians are urged to leave behind in the first oracle *before* Xerxes attacks might actually apply, as Macan has observed, more suitably to Athens and Attica *after* the occupation.[23] Similarly, the reference to the city as "wheel-shaped" (τροχοειδέος) better describes Athens after the Themistoclean walls had been built rather than the unwalled city at the time the Persians attacked, underscoring

22 Cf. Maurizio 1997: 318 who underscores the "fluidity between past, present and future" in the Delphic narrative: "Past utterances become understood in the present because they are reconstructed through the present circumstances. Thus, oracular performances replicate the permeability between different times, as oracles themselves as prophetic utterances do . . ." Maurizio goes on to note that Herodotus' text "fossilizes both the relationship between past and present and the fluidity that attended any oracular performance."

23 Macan (1908) 1973 at Hdt.7.140 notes the similarities between the account of devastation alluded to in the language of the oracle and Herodotus' description of the subsequent sack of Athens in Book 8.50–53. I don't agree, however, with Macan's conclusion that this is evidence that the oracle is a later forgery.

the extent to which Athens' city walls have come to symbolize the city.[24] The city that Herodotus' Delphic narrative constructs, then, is not the historical city of 480 as Maurizio argues, but rather a composite Athens, one that includes all the key elements that have come to define the city of Athens differently at different times – walls, land, ships, inhabitants – in the fifty years after Salamis.[25] In addition to offering advice to the Athenians about what to *do* about their city, the oracle also suggests the terms for how to *think* about their city, newly configured both physically and economically as a naval empire.[26] In choosing to structure his narrative of the events prior to Salamis as a Delphic prophecy to be interpreted then and now, Herodotus thus gives his audience both an account of what happened at that critical moment in 480 and the framework for returning again and again to the basic civic elements (ships, walls, men) that continue to define their city even as it has

24 Harrison 1981. In fact, in an article discussing the iconography of Athena's shield, Harrison (p. 85n94) draws upon the language of the Herodotean oracle as part of her analysis of the shield's topography: "The image of the city in this oracle given to the Athenians, 'the houses and lofty top of your wheelshaped city,' is something like that on the Shield."

25 In spite of her sensitivity to the orality of the Delphic tradition, Maurizio 1997: 331 reads the oracles' interest in the definition of the city of Athens as a product of the 480s: "Both oracles indirectly revolve around this question, one that we can imagine would naturally have presented itself at this time in Athens' history, a time when Greece's identity has come into question in the face of the Persian invasion, and individual Greek cities must define themselves in relation to Persia and to one another."

26 Indeed, recent scholarship confirms this reading of the two oracles together by emphasizing the role that Herodotus plays in constructing as well as recording the tradition of Delphi as ambiguous within his *Histories*. Barker 2006: 23 argues that Herodotus uses oracular consultation as a historiographical strategy that constructs his own text as an arena of interpretation: "his re-presentation of the two oracles invites a reader *to participate fully in the processes of decision-making*" (his italics). Barker goes on to say: "It is not just the oracle that is the site of struggle and conflict; Herodotus' text itself becomes the battleground over which the readers fight for control of meaning and for their own identity."

changed from a polity that prized its autonomous and autochthonous agrarian roots to one that looks to ships and the sea to define its civic identity.

Sophocles' *Oedipus the King*: "Neither a Walled Town nor a Ship"

I want to turn now to Sophocles' *Oedipus the King*, a play that, in spite of its Theban setting and plot, scholars have often read productively for its engagement with contemporary Athenian politics.[27] We need not go as far as Knox's allegorical reading of Oedipus as imperial Athens to recognize and explore the resonance that a play about Thebes' civic crisis might have had for an Athenian audience fresh from the devastations of both the plague and the renewed hostilities with Sparta. While Sophocles does not refer to the "Wooden Wall" oracle specifically, he does combine the iconic themes of ships and city walls in a way that recalls the competing civic terms contained within the oracle and the story of its decipherment. Indeed, almost forty years earlier, Aeschylus invoked this same thematic binary of ships and walls as part of his own staging of the Theban saga in the *Seven Against Thebes*.[28] In this play, produced in 467 BCE, Aeschylus dramatizes the fratricidal conflict between Oedipus' two sons, Polyneices and Eteocles, that takes place at one of the walled city's famous seven gates. In this context, Aeschylus represents the danger inherent in this particular moment in Thebes' extended civic crisis by bringing nautical imagery together with the figure of the city wall.

27 Most famously, Knox 1957; more recently, see also Mitchell-Boyask 2008: 56–66; Vickers 2008. Cf. Zeitlin 1990 for the argument that Athenian dramatists consistently constructed Thebes on the Attic stage as the "anti-Athens."

28 Those who have argued for the discursive power of Thebes on the Athenian stage (e.g., Zeitlin 1990; Rader 2009) have tended to focus on autochthony and its differential political associations in Athens or Thebes, but Thebes' iconic status as a walled city may be equally productive for Athenians coming to grips with the implications of Athens' growing urbanization in second half of the fifth century.

As the Argives threaten to attack, the chorus worries that the city's prow is at risk of sinking and that its walls will not prove strong enough (758–65). And yet, in spite of the threat posed by Polyneices and the Argive army, the messenger reports that the city has survived the attack:

> πόλις δ' ἐν εὐδίᾳ τε καὶ κλυδωνίου
> πολλαῖς πληγαῖς ἄντλον οὐκ ἐδέξατο.
> στέγει δὲ πύργος, καὶ πύλας φερεγγύοις
> ἐφραξάμεσθα μονομάχοισι προστάταις·
>
> Our city is in smooth water and even with many
> assaults of the waves, has not taken on water.
> Our wall protects us, and our gates
> we fortified with capable protectors fighting in single combat. (Aesch. *Sept.* 795–98)

Aeschylus here calls upon both the ship of state image and the figure of the city walls to represent Thebes' survival at the brink of destruction. Both images work together cooperatively rather than competitively (as they did in Herodotus' account of the Persian invasion) to celebrate the city's survival in terms of its infrastructure.[29] In Aeschylus' play, the city is defined by its infrastructure – whether that be figured as a ship at sea or in terms of its city walls – and celebrated for its ability to protect its citizens against an external threat.[30]

Sophocles, writing *Oedipus the King* nearly forty years later, returns to Thebes, then, to revisit Aeschylus' optimistic faith in the city's infrastructure by staging an earlier point in the Theban myth to imagine a more problematic relationship between a city's people and its infrastructure faced with a

29 Cf. Goldhill 2007 who discusses the wall imagery in this play as well as in Euripides' *Phoinissae*. Goldhill's discussion focuses primarily on some instructive differences between the ways in which each poet extends the Homeric tradition and uses wall imagery to engage with issues of personal and political identity as well as the relationship between tradition and political behavior.

30 Aeschylus here inverts the analogy he used in the *Persians'* passage discussed above. There the survival of the men means the survival of the city's walls; here, the walls' survival means that the city and its men are safe.

different kind of civic challenge.[31] In this play, I will argue that Oedipus, a mythic figure famous for his skill at solving the riddle of the Sphinx, is called upon to help the Athenians think about the challenges facing their own city a half-century after solving the riddle of the "Wooden Wall" oracle, a city now firmly configured as a city of both walls and ships.

As the play begins, a priest comes to ask Oedipus for help since the city of Thebes is suffering from a devastating plague. Dwindling crops and flocks, stillborn children, and a "most hostile plague" (λοιμὸς ἔχθιστος, 28) descend upon the citizens of Thebes – language and imagery that could not help but recall the recent experience of Sophocles' Athenian audience with the plague that devastated Athens in 430.[32] The priest warns Oedipus that it is his responsibility to take steps to protect the citizens of Thebes. After all, he saved the city once before from the monstrous Sphinx, and the priest concludes his appeal with a familiar metaphor, encouraging Oedipus to help the citizens of Thebes:

> ὡς εἴπερ ἄρξεις τῆσδε γῆς ὥσπερ κρατεῖς,
> ξὺν ἀνδράσιν κάλλιον ἢ κενῆς κρατεῖν·
> ὡς οὐδέν ἐστιν οὔτε πύργος οὔτε ναῦς
> ἔρημος ἀνδρῶν μὴ ξυνοικούντων ἔσω.

> For if you wish to rule this land, just as you hold power,
> It is better to hold power with men rather than over an empty land;

[31] Euripides' *Phoinissae*, then returns yet again to the Theban story; this play puts the city walls in the middle of a much more permeable interpretation of walls as constitutive of the city as Goldhill 2007 has noted.

[32] There is no secure date for the play, but I am persuaded by Knox 1956 who dates the play to 425 BCE; he is followed by Kamerbeek 1967: 28–29, although with more uncertainty, and Mitchell-Boyask 2008: 56–66, who discusses the theme of the plague in the play. Dawe 1982: 245 rejects Knox's argument for a 425 date as brilliant but unconvincing; Müller 1984: 59 argues against Knox and for a date of 433 for the play. For arguments for reading the plague in the *Oedipus the King* as a reference to Athens' recent plague, see Knox 1956; Kamerbeek 1967: 38; Parry 1969: 114.

For neither a walled city nor a ship is anything
if empty of men dwelling within it. (Soph. *OT*, 54–57)[33]

In this passage, the priest invokes both of the key civic models that are debated in Herodotus' account of the "Wooden Wall" oracle and combined in Aeschylus' *Seven*. The choice between the defensive strength of city walls and the aggressive potential of a navy so much at issue in Herodotus' second oracle, however, is here rendered moot by the functional equivalence of city walls and ship – neither is worth ruling, the priest contends, if empty of men (ἔρημος ἀνδρῶν). Here the infrastructural choice between ships and walls is irrelevant, and in its place Oedipus must choose between a city with or without men.[34] Sophocles' image calls attention to the city's population, referring to men at the beginning of the passage (ξὺν ἀνδράσιν) only to emphasize their disappearance through the repetition of words for emptiness (κενῆς, ἔρημος). The final genitive absolute (ἀνδρῶν μὴ ξυνοικούντων), in fact, reverses the emphasis on the sufficiency of the strength of men to define Athens that we saw in the passage from Aeschylus' *Persians*.

Sophocles thus challenges the assumptions behind the competing visions of the city as they are articulated in the two Herodotean oracles by raising questions about the sufficiency of any kind of infrastructure to represent the city. In this play, the very ships and walls, which in the combined vision of Themistocles and Pericles have come to represent Athens' newfound identity as a naval and imperial city, are now shown to be insufficient symbols of a city's identity. In the absence of infrastructure, the defining element of the city becomes its

[33] Cf. 1378 for the collocation of wall or tower (πύργος) with city (ἄστυ). See also Soph. *Ant.* 953 for the collocation of walls and ships: οὐ πύργος, οὐχ ἁλίκτυποι/ κελαιναὶ νᾶες. Sophocles' image here also reminds us very much of the Alcaeus passage cited earlier. Earle 1901: 149 explains that πύργος equals τεῖχος and that the wall stands for the city. He also notes, "Sophocles is probably thinking of the Long Walls of Athens."

[34] It is significant, here, that the city is still configured in terms of ships and walls, even as its existence is dismissed.

inhabitants and their ability to animate the city (as we saw in our reading of Herodotus' first oracle.) Indeed, there are interesting linguistic similarities between Sophocles' description of the city of plague-ridden Thebes and the image of animated buildings in distress in the first oracle recounted by Herodotus.[35] Through the voice of the priest, Sophocles thus puts before Oedipus a puzzle that, I contend, would be familiar to an Athenian audience in that it contains precisely the same civic debate over infrastructure contained within Athens' famous "Wooden Wall" oracle, as well as a framework – albeit a very different one – for trying to solve it.

Sophocles' representation of Oedipus thus functions as a kind of mythical analogue to Herodotus' deployment of Themistocles and the Delphic oracle. Not only do Herodotus and Sophocles deploy the prophetic skills of Themistocles and Oedipus similarly, there are connections between these two texts and Thucydides' account of yet another situation of civic distress, namely the Athenian plague. Scholars have already noted the linguistic similarities between Sophocles' play and Thucydides' account of the plague, but now that we have established a commonality between Sophocles' and Herodotus' texts, a coherent pattern of imagery used to represent a city in crisis from within emerges from reading all three together.[36]

35 Herodotus' oracle warns that "neither the head remains intact, nor the body, nor the hands, feet or middle" (οὔτε γὰρ ἡ κεφαλὴ μένει ἔμπεδον οὔτε τὸ σῶμα / οὔτε πόδες νέατοι οὔτ' ὦν χέρες, 7.140.2). The oracle depends upon corporeal imagery very much like the language with which the priest in the *OT* warns that the city is floundering as if in a storm at sea and can no longer hold up its head (22–24). In addition, the animated Athenian civic buildings in the Herodotean oracle shake with fear (δείματι παλλόμενοι) as does the opening chorus of the Sophoclean play (δείματι πάλλων, 153). In Sophocles' play, as in Herodotus' text, the animating metaphor serves to emphasize the weakness rather than strength of the city.

36 For similarities between Sophocles' text here and Thucydides' account of the plague, see Shephard 1920: 100; Knox 1956; Parry 1969. See Bushnell 1988: 69–72, who reads the *Oedipus the King*'s preoccupation with prophecy together with Herodotus' account of Themistocles and the wooden wall oracle.

In addition to "predicting" the destruction of Athens by the Persians, Herodotus' account of the first oracle, with its sweating and shaking temples, may also "foretell" an internal threat to the integrity of the city – the plague – one that follows from its embrace of walls and an urban civic identity. Certainly, as Thucydides makes clear in his account (2.52), the sufferings and deaths from the plague in Athens were exacerbated by Athens' recent trend toward increased urbanization. Similarly, Sophocles' dramatization of the plague in Thebes on the Attic stage in the years immediately after Athenians had suffered their own plague can be read productively alongside the "Wooden Wall" oracle as part of an extended meditation on the Athenian city in a different kind of crisis – this time from within – one that raises real concerns about the shape of the city at this point in time. This is not to say that any one of these three authors is drawing directly upon another's text, but rather that all three participate in a shared vision of the city and the challenges it faces.

Sophocles taps into the incestuous duality at the core of Oedipus' mythic persona to think about ships and walls together as a figure for the city in rather more pessimistic terms than does Herodotus. What does it mean to be a city of ships *and* walls? Over the course of the play, Oedipus himself is compared to both the ship of state and the city wall. While Sophocles draws upon "ship of state" imagery to describe the fragility rather than the resilience of Thebes' civic affairs (*OT* 4, 22–24), Oedipus is invoked, at the beginning of the play at least, as the competent helmsman who can right the ship. As the play opens, Creon greets Oedipus as the one who has steered a straight course for the city (σὲ τήνδ' ἀπευθύνειν πόλιν, 103–04) through its recent turbulent times much as Solon is encouraged to do by the Delphic oracle (εὐθύνων).[37] And yet, once

[37] Later on in the play, the chorus praises Oedipus' political skills with imagery drawn from sailing (694–96), and Jocasta, as well, boasts that all of Thebes looks to Oedipus as "the steersman of the ship (of state)" (ὡς κυβερνήτην νεώς, 923). Cf. the oracle allegedly received by Solon (Plut. *Solon* 14.6) cited above.

Oedipus' true identity is known and the story of Thebes' tragic events unfolds, his role as competent captain of the ship of state disintegrates and merges into that of the troubled ship itself. Indeed, the play concludes with a lament over the depths to which Oedipus has fallen, again phrased in nautical terms: "he has come to such a shipwreck of terrible misfortune" (εἰς ὅσον κλύδωνα δεινῆς συμφορᾶς ἐλήλυθεν, 1527). Oedipus has become the ship of state, and this collapse of terms bodes ominously for the future of Thebes.[38] Whereas the association of ship and state was celebrated by Herodotus' Themistocles as the means of saving the city, here Sophocles suggests that the city figured in this way is in deep trouble.[39]

Moreover, in addition to personifying the troubled Theban ship of state, Sophocles' Oedipus is also aligned with the city's defensive structures (πύργος) toward the end of the play. As the chorus attempts to grapple with its new knowledge of Oedipus' true identity, they praise Oedipus for conquering the Sphinx, calling him a bulwark against death for the city (θανάτων δ' ἐμᾷ/ χώρᾳ πύργος ἀνέστας, 1200–01), thus invoking him as a great king of Thebes with a choice of language that recalls Alcaeus' formulation of men as the walls of the city (πόλιος πύργος).[40] Oedipus is the bulwark (πύργος), the one who defends the city from external threats, and he is the steersman of the ship of state (κυβερνήτην νεώς), capable of guiding his city with

38 In fact, Sophocles merges the ship of state imagery with the incestuous acts at the core of Oedipus' story in devastating terms as we see when the chorus draw once more upon ship imagery to represent Jocasta's womb as a harbor: "The same great harbor sufficed to fall for son and/father/bridegroom" (1208–10). Cf. Teiresias' veiled prediction of Oedipus' future, also expressed in these terms (420–23).

39 For a list of all the uses of the ship imagery in this play, see Campbell 1986. In this context, the conventional ship of state imagery takes on new meaning. Similarly, the play's obsession with agricultural imagery – applied both to Oedipus himself, the man who sowed seeds from the same field from which he had sprung, and to the city, whose crops and produce have dwindled and failed – recalls the notion expressed in Herodotus' first oracle, that the city equals the land, expressed here in agricultural, rather than in terms of buildings. Cf. 1211–12; 1482; 1496–99.

40 See also Aesch. *Sept.* 798.

a fair wind at its back. But while Aeschylus' *Seven* productively combines ships and walls as complementary metaphors for the city, metaphors that emphasize their ability to save its citizens in times of distress, here, precisely because the duality of ships and walls is triangulated through the ambiguous figure of Oedipus, the combination is presented as a problem rather than a source of protection for the city.

In his ground-breaking reading of the play, Jean-Pierre Vernant emphasizes the inherent duality that governs Sophocles' representation of Oedipus: "Like his own discourse, like the pronouncement of the oracle, Oedipus is double, enigmatic.... the figure of Oedipus proves to be in every aspect – social, religious, and human – the opposite of what, as leader of the city, he seems to be."[41] Just as Oedipus is both leader and scapegoat, native and foreign, healer and diseased, he is also associated with both competing models for the city in Herodotus' account of the wooden wall oracle: city as ship versus city as defensive walls. In other words, more than a good riddle solver (as was Herodotus' Themistocles), Oedipus embodies the riddle of the city itself, and the incestuous plot at the heart of Sophocles' play suggests that what saved the city once, namely the interpretation of the "wooden wall" as ships, may no longer be the right choice, especially if it then leads to the emergence of Athens as a city of walls as well. How can either defend a city against an internal threat? Rather than celebrating the city's newfound imperial strength as a city of ships and walls, Oedipus' ultimate downfall at the end of the play raises serious questions about the city's infrastructure – about the choices it has made in the past and its ability to make good (or better) choices for its future.[42] As the

41 Vernant 1988: 118–19; the quotation continues: "The Corinthian stranger is in reality a native of Thebes; the solver of riddles is a riddle he himself cannot solve; the dispenser of justice is a criminal; the clairvoyant, a blind man; the savior of the town, its doom."

42 Athens' decision to bring all those living in the agricultural outskirts of town inside the Long Walls in the face of hostile attacks by the Spartans was, of course, a key factor in increasing the devastation brought by the plague; cf. Thuc. 2.52.

priest points out to Oedipus at the beginning of the play, it is not worth ruling a walled city or a ship if they are empty of men (this, after all, is the deserted Athenian city the Persians captured). Is Oedipus' immanent exile a charge to Athenians to reject similarly the formulation of walls and ships as constitutive of the city? Or does Oedipus' future transformation from defiled scapegoat into powerful civic hero, one that takes place beyond the scope of this play but which is certainly relevant to the well-being of Athens, hold out hope that Athens, too, may be similarly transformed? No matter how the end of the play is interpreted, it is clear that rather than focus on the choice between ships and walls, as did Herodotus' narrative, Sophocles' play explores the implications of their inextricability, emphasizing the dangers that befall a city that has come to be defined, indeed confined, by both ships and walls.

Thucydides' Sicilian Narrative: "Men are the City"

I want to conclude this discussion of the engagement between Athens' men and its infrastructure in the city's civic imaginary by turning to Thucydides' account of the Sicilian expedition, written with full knowledge of Athens' loss to the Spartans in the Peloponnesian war – a loss that was, in fact, symbolized by the destruction of the city's ships and walls.[43] Scholars have already identified the prominent role that ships and walls play in the historical analysis of power outlined in Thucydides' Archeology in book 1 of his *History*, both generally and with

43 At 5.26.1, a passage that has been called the "second preface," Thucydides describes the end of the Athenian empire as the capture of the city's Long Walls and the Piraeus. There is no definitive date of publication for Thucydides' text nor consensus on its mode of composition. In his introduction to the *Landmark Thucydides*, Hanson 1996: xiii suggests that Thucydides wrote and revised his *History* on and off from 431 to the early 390s when his death cut short his narrative at the events of 411. Hornblower 1996: 119–22 reluctantly reviews arguments for different modes of composition over time, suggesting that "one might want to avoid talk of publication date

respect to Athens' own imperial aspirations in the fifty years after the Persian invasion.[44] Thucydides returns to these themes in book 6 as he introduces the Sicilian expedition, in a passage that serves, as W. R. Connor notes, essentially as a second Archeology. Noting the correlations between books 1 and 6, Connor argues that events surrounding the Sicilian expedition forced Thucydides to reconsider his analysis of power as outlined in book 1, and it is in this context that I would like to turn now to Thucydides' representation of ships and walls in the Sicilian narrative, to discuss the poetics rather than the realia of Athenian civic infrastructure in this text.[45]

Thucydides' account of the Sicilian expedition, as it unfolds in books 6 and 7, deploys a variety of narrative patterns familiar to his Athenian audience. First, he introduces the Athenian expedition against Sicily immediately after his powerful description and searing analysis of the Athenian siege of Melos. This abrupt narrative juxtaposition of city at siege and naval expedition thus gives the reader an ominous sense of foreboding about the future of this naval expedition as well.[46] Second, as many scholars have shown, Thucydides structures his account of the Athenian expedition against Syracuse as the

for Thucydides." Cf. Dewald 2005: 6 for the dangers of linking discussion of composition date with stylistic analysis.

44 Cf. Hedrick 1995: 76–86; Ober 1985; Connor 1984: 23–27.

45 Connor 1984: 160: "The Archaeology, as we have seen, was a confident analysis of power, emphasizing the advantage enjoyed by naval and imperial states over continental land powers.... In the Sicilian narrative quite a different analysis of power is developed. The larger pattern of the sixth and seventh books, in which not only Athenian power but even the categories of analysis of power are confounded, is anticipated in the opening chapters of the sixth book."

46 Cf. Connor 1984: 158 and note 1 for Thucydides' careful juxtaposition of the Melos and Sicilian narratives. See also p.161: "Even at the outset, however, the reader is aware that what is in store for the Athenians is the greatest single disaster of the war. The prospect, moreover, is not entirely unwelcome. Melos is too fresh in our minds and the Athenian severity too unmitigated. Such violence invited retribution. In the old Greek mythic world view the excesses of conquerors were often punished during their return over the high seas."

inversion of Athens' famed naval success at Salamis.[47] Once again, the Athenians attempt to save their city by leaving their city, but this time Athens occupies the role of the bloated and overly confident imperial force that is brought low by the scrappy, underdog Sicilians. Thucydides' Sicilian account looks back to the iconic Salamis narrative, to a time before the city had real walls and relied upon their fleet, as a framework for assessing the nature of the Athenian city fifty years later – a city now defined by its walls and ships.[48] Finally, Thucydides infuses his Sicilian narrative with themes familiar from archaic Greek colonial narratives, thus representing the aggressive military expedition as a kind of civic renewal, an attempt on the part of the Athenians to save their city at home by refounding it in Sicily.[49] By drawing upon this rich tapestry of familiar narrative patterns, Thucydides makes his description of the Sicilian expedition clearly relevant to the Athenian experience at home – past, present, and future. He weaves together fond memories of the city's finest moment, ominous anxiety about the success of the expedition, and renewed hope for Athens' future. In addition, the Athenian expedition against Syracuse was, as Thucydides tells it, a battle of both ships and walls, and in this respect, his narrative engages directly with these key infrastructural elements of the Athenian civic imaginary.

Right from the beginning of his Sicilian narrative, Thucydides elaborates the identification of Athenian civic identity with its fleet, a notion that goes back to his account of Athens' decision to follow through on Themistocles' advice that the Athenians take steps to "become naval" (ναυτικοὶ ἐγένοντο, 1.18.2; cf. 1.93.3) as well as to Pericles' exhortation to "safeguard the city and the sea" (τῆς δὲ θαλάσσης καὶ

47 Cf. Connor 1984: 175; Rood 1999; Harrison 2000; Taylor 2010: 138.n8.

48 Cf. Connor 1984: 176: "As the city of Athens moves across moral and physical boundaries in its attack on Sicily, it is also moving chronologically backward to confront its own past."

49 Cf. Avery 1973; Taylor 2010: 146. Cf. Dougherty 1993: 15–30 for discussion of the kinds of narrative patterns that define the colonial narrative.

πόλεως φυλακὴν ἔχειν, 1.143.5). In a speech delivered after the Athenians have already voted to sail to Sicily, Nicias, one of the Athenian generals charged with leading the expedition, tries to persuade his fellow Athenians to change their minds, warning them against indulging in ever greater imperial aspirations while the state of their current empire was under threat, and he does so by linking the Athenian city with its ships. He reminds them that

> it is necessary to consider all these things, and it is not, when the city is still at sea (μετεώρῳ), worthwhile to run risks or to grasp after another *arche* before we have strengthened the one we already have. (Thuc. 6.10.5)

Although the word that Nicias uses here (μετεώρῳ) literally refers to things "in the air," elsewhere Thucydides regularly uses the same word metaphorically to describe a ship on the high seas (e.g., the Corinthians "drew up their ships on the open sea but remained quiet" [παραταξάμενοι μετεώρους ἡσύχαζον, 1.52.2]; "a merchant ship in open water" [ὁρμοῦσα μετέωρος, 2.91.3]).[50] In encouraging his fellow Athenians to take care of matters at home before venturing after more conquests abroad, Nicias draws upon the familiar ship of state image, one that, as we have seen, has also taken on a particular set of associations given the specific experiences of Athenians in the fifty years since the battle of Salamis. Nicias' evocation of the ship of state to describe Athens, poised on the brink of the Sicilian expedition, extends this image in ways that combine both the Herodotean and Sophoclean examples. For Nicias, as for Herodotus' Themistocles, the Athenian

[50] Cf. Taylor 2010: 140–3; on p. 141 she lists for these and additional examples. Cf. Hornblower 2008 *ad loc.* who translates this passage as "the city is far from dry land." In addition to noting Thucydides' customary use of this word to refer to ships out at sea, he cites a scholiast who paraphrases the passages as "when our city is not safely anchored." While acknowledging that the word may have a medical flavor, "in a delicate position," he notes that both are in play, observing that "the meaning of key Thucydidean terms is sometimes unstable." Cf. Gomme 1970 *ad loc.*

state is captured metaphorically as a ship, but by contrast to Themistocles' two hundred fully-manned ships confidently sailing off to meet the Persians at Salamis, this ship of state is "up in the air"; it is on the high seas, unmoored. While the ship is not in immediate danger of disaster as is Sophocles', it is far from shore and safe harbor, and its future safety is far from secure. In this way, Nicias' image underscores the unresolved nature of the Athenian ship of state at this point in time, and for this reason, as Nicias argues, it should not take on additional risks.[51]

Unable to dissuade his fellow Athenians from setting sail to Sicily, however, Thucydides' Nicias tries again. In a second speech, he painstakingly lays out the dangers and difficulties of such a large expedition, and to do so he compares the military expedition to an overseas colonial endeavor:

> We must consider that we are going to found a city among foreign and hostile people and that we must control the country land straightaway on the first day we occupy it or know that if we fail, everything will be hostile. (Thuc. 6.23.2)[52]

In addition to highlighting the danger and preparations necessary for such an expedition as in his previous speech, here, Nicias' choice of colonial imagery conjures up the Sicilian expedition as a kind of city on a ship, one that he imagines to be heading for Sicily.[53]

51 Taylor 2010: 143. As Taylor suggests, "Nicias' image of the city on the eve of the Sicilian Expedition, then is of a ship at sea, running for harbor from the dangers of the ocean."

52 In this respect, Nicias' words are a conflation of the two choices that Herodotus' Themistocles poses: either attack the Persians at sea or sail to Sicily to found a new city (cf. Hdt. 8.61–62). Here they are merged into one.

53 Taylor 2010: 145: "With these words Nicias imagines the Sicilian expedition as a floating city … as a city at sea – the very thing that, he had used in his earlier speech as an image to scare the Athenians into voting against the campaign. The *meteoros* polis, which Nicias had originally urged the Athenians to ground in order to secure the old empire, now has a new destination."

The Athenians ignore Nicias' advice, and Thucydides' use of the colonial image in this speech represents the Athenian decision to attack Syracuse in interesting terms. First, the conflation of the Salamis and colonial narratives suggests that instead of leaving their city behind and taking to their ships, as they did at Salamis, the Athenians have chosen to take the city with them. Second, Thucydides' formulation represents the Athenian decision to attack Syracuse as if they were attempting to create a second Athens, one that is both enabled by and embodied in their ships.[54] The Salamis narrative looks back toward the city left behind while the colonial story focuses on what lies ahead, the city still to be founded. By contrast, Nicias' ship of state, at the beginning of the Syracuse story at least, is stuck in the middle (of the sea and the story), unmoored, unclear about where it will end up. As Thucydides tells it, the Athenians leave Athens (Salamis narrative) as if to found a new Athens (colonial narrative) which in turn they must abandon (Salamis narrative) to return to Athens – to refound it (colonial narrative)? To start anew? Does Thucydides' repetition of these narrative patterns offer hope for a renewed Athenian city? Or does it rebuke the Athenians for believing in that same old story – that they can continue to save their city by abandoning it?

Walls, the other half of the civic dyad we have been exploring, are also a key element in Thucydides' narrative of the Sicilian expedition. At the end of book 6, Thucydides embarks on a lively account of what W. R. Connor has dubbed the "race of walls" between the Athenians and the Syracusans, one in which army and city, walls and ships, Athens and Syracuse are purposefully conflated.[55] At 6.98.2,

[54] Thucydides' Sicilian narrative thus offers a rich and allusive set of notions that are all tied up in his deployment of the ship as a metaphor for the city of Athens. Herodotus' Salamis narrative of a heroic defense of Athens has become a more aggressive, imperial/colonial tale, but the identification of the city with its ships remains constant.

[55] Connor 1984: 187.

Thucydides observes that the Athenians "advanced to Syca, where they halted and quickly built the circle, or center wall of circumvallation" (ἵναπερ καθεζόμενοι ἐτείχισαν τὸν κύκλον διὰ τάχους). The account that follows dramatically details the back and forth, Athenians and Syracusans, walling (τειχίζειν) and counterwalling (ὑποτειχίζειν). It ends with the dramatic midnight efforts on the part of the Syracusans to successfully outstrip the Athenians and establish a counterwall that would effectively prevent the Athenians from walling off their city (7.6.4), thus halting the Athenian land offensive and foreshadowing the similarly unsuccessful results of the Athenian naval campaign.[56] Indeed, Thucydides has temporarily postponed the account of the naval expedition and replaced it with this wall-building narrative, a discursive move that underscores the connections between walls, ships, and cities in Thucydides' understanding of the Athenian civic experience.[57]

In addition, Thucydides' attention to walls as a fundamental part of the Athenian civic imaginary exploits the inherent ambiguity, what we might call the "two-sidedness" of walls. While Sophocles' Oedipus embodied the ambiguity inherent in representing the city as both ship and walls, Thucydides suggests that walls themselves are an ambiguous figure for the city. Do they wall something in or keep it out? Are they tools of aggressive siege warfare or the protective symbols of a new city? First, Thucydides represents Athenian efforts to build siege walls that would cut Syracuse off from its supplies

56 Thucydides' account of this "race of walls" runs from 6.98–7.6.4, where he notes that Syracuse successfully completed their counterwall, depriving the Athenians of any chance to wall off their city (ἀποτειχίσαι).

57 Hornblower 2008: 551–54 notes Thucycides' brilliant and innovative use of vocabulary to represent the building of walls and counterwalls by both Athenians and Syracusans. He argues for Thucydides' "stylistic enactment" of the race of walls between Athens and Syracuse by noting the "unusual piling up of heavy building words" and is surely right to underscore the narrative significance of Thucydides' account of the wall building within the larger Sicilian narrative.

as if the Athenians were building a city, one that looks suspiciously like the Athens they left at home.[58] And, then, just as at home, he describes the Athenian soldiers as the ones besieged.[59] As Nicias explains in a letter back to Athens, because the Syracusans were able to foil Athens' strategy through their counter wall, "it turned out that the one who seemed to be besieging, suffered that experience" (ξυμβέβηκέ τε πολιορκεῖν δοκοῦντας ἡμᾶς ἄλλους αὐτοὺς μᾶλλον, ὅσα γε κατὰ γῆν, τοῦτο πάσχειν, 7.11.4). The inherent ambiguity about walls is highlighted as "those who seemed to be doing the besieging" (πολιορκεῖν δοκοῦντας) turn out to be the ones being besieged. The only real difference is which side of the wall they are on.

The proliferation of competing narrative patterns at work (siege, evacuation, colonization) in Thucydides' Sicilian narrative helps exploit this reflexivity or ambiguity inherent in walls and raises interesting questions about their stability as core elements of a city's infrastructure. Are these walls like

58 Walls, of course, are among the first things built by new colonists to mark the boundaries of their territory, as Homer tells us at *Od* 6.7–10. Toward the end of book 6 (6.103.1), Thucydides mentions that the Athenians are building a double wall (τείχει διπλῷ) from Epipolae down to the sea, and Avery (1973: 12) observes that "the double walls represent a sort of miniature version of the long walls reaching from Athens to the sea at the Piraeus and Phaleron. If this were so it would help support a sense of identity between the army abroad and the city at home."

59 For the city of Athens, itself, has essentially come under siege from the Spartans, and Thucydides shifts his narrative from the Syracusan front to events back in Athens where he observes that the Spartans had successfully fortified Decelea, thus cutting Athens off from its countryside. As a result, Thucydides explains, "the city needed every single thing to be imported and instead of a city it had become a fortress" (τῶν τε πάντων ὁμοίως ἐπακτῶν ἐδεῖτο ἡ πόλις, καὶ ἀντὶ τοῦ πόλις εἶναι φρούριον κατέστη 7.28.1). Thucydides further articulates the connections between the Athenian experience in Athens and that of the army in Sicily at 7.28.3 and 7.75.5, where he describes the retreating Athenian army in Sicily as "nothing other than a besieged city in flight, and not a small one" (ουδὲν γὰρ ἄλλο ἢ πόλει ἐκπεπολιορκημένῃ ἐῴκεσαν ὑποφευγούσῃ καὶ ταύτῃ οὐ σμικρᾷ,).

the walls of Troy, walls that confine the citizens within and lay them open to siege? Are they a mirror of Athens' own city walls, destroyed by the Persians in 480? Or are they the walls that colonists build immediately upon founding a new city, a defining marker of its existence? Thucydides' account explores all these possibilities, underscoring the instability of walls as secure symbols of civic identity at this point in the Athenian experience.[60]

At Syracuse, both the naval and wall offensives fail, of course, and fail disastrously, and Thucydides recounts the Athenian losses in great detail, culminating with Nicias' valiant but futile attempt to encourage his routed troops in the face of utter defeat. Attempting to restore some hope to his defeated men, Nicias says, invoking what should by now be a familiar collocation, "Men are the city and not walls or ships empty of men" (ἄνδρες γὰρ πόλις, καὶ οὐ τείχη οὐδὲ νῆες ἀνδρῶν κεναί, 7.77.7).[61]

The opening of Nicias' formulation, "men are the city" (ἄνδρες γὰρ πόλις) takes us back to the Alcaeus fragment (ἄνδρες γὰρ πόλιος πύργος ἀρεύιοι) which we discussed earlier – but with one conspicuous difference. In Alcaeus' words, men are associated with the city through the mediating image of the city's walls; men are strong like the walls and together they encircle and protect the city. As we noted earlier, the word order of the Alcaeus fragment emphasizes the location of the city within the walls, defined by the two words for men. In Nicias' formulation at the conclusion of Thucydides'

60 On the narrative complexity of Thucydides' Sicilian account, with particular attention to the ambiguity he exploits between the colonial and evacuation narratives, see Harrison 2000:93–94. Bassi 2007: 190 also identifies a key link between these two defining themes in Athenian civic imaginary: "In short, the Athenians' famed willingness to leave their city behind – including the idealization of the city that results – may be understood as the necessary antidote to being confined within it."

61 This is a much discussed passage, see Longo 1975: 78; Connor 1984: 202–03; Croally 1994: 168; Harrison 2000: 94; Bassi 2007: 190–97; Hornblower 2008 *ad loc.*

Sicilian narrative, however, the very same word placement signals something very different: ἄνδρες γὰρ πόλις, καὶ οὐ τείχη οὐδὲ νῆες ἀνδρῶν κεναί. In Nicias' chiastic formulation, the words for men (ἄνδρες, ἀνδρῶν) are now located outside the city, defined here as the absence of ships and walls, key infrastructural elements (πόλις, καὶ οὐ τείχη οὐδὲ νῆες), and this displacement of the city's population is reinforced by the postponement of the word for empty (κεναί). Whereas Alcaeus' men were strong like city walls, a metaphor that emphasizes civic strength, here the ships and walls are empty and thereby revealed as deeply flawed metaphors for the city. At the end of the sentence (as at the end of the story), the city is not *strong like* men, but *empty of* men. The resulting image simultaneously captures the routing of Athenian troops in Sicily and the abandonment of the Athenian city in favor of the Sicilian campaign, undermining the fundamental role that ships and walls played in both. Nicias' plea reminds us that while in Herodotus' Salamis narrative, Xerxes' capture of an empty city meant the survival of Athens, here it foretells its destruction.[62]

Nicias' final formulation thus reverses the emphasis that Thucydides made in book 1 on infrastructure – on ships and walls – that was so important to Athens in the years immediately following the Persian wars. As Connor notes, Thucydides inverts "the calculus of power in the archaeology with its emphasis on the physical and quantitative bases of power, especially walls and ships. In the last ironic analysis all depends on men, not on material resources."[63] Nicias' metaphor that defined the city as the naval expedition,

62 In addition, as Bassi 2007: 191 observes, Nicias' formulation goes even one step further to move the Athenian city out of Athens: "he equates Athens not with the men who dwell within it but with its troops in the field. In fact, the metonymic effect of his statement depends on the fact that the troops are not in the city or rather, that the city is an empty space because of their (the troops') absence."

63 Connor 1984: 202–03.

with all its connections between ships and walls, has been stripped of the terms of comparison, leaving nothing, not even the men themselves.

CONCLUSION: SHIPS, WALLS, MEN

A close reading of selected passages from these texts reveals a sustained engagement with the conceptual potential of ships and walls as defining components of the city's infrastructure. It is precisely because they are a critical part of the lived experience of Athenians in the fifth century that ships and walls are such powerful conceptual tools, and this paper set out to explore the place where historical reality engages with discursive potential in an effort to see what can be learned from elaborating such a poetics of infrastructure.

The ship, of course, represents Athens' recent emergence as an imperial power and a decision to privilege mobility over the city's settled autochthonous past. As we've seen, ships play a key role in a variety of narratives important to Athens' recent history, whether they are tales of escape or conquest, unanticipated victory or unparalleled defeat.[64] And yet, precisely because in the years between 480 and 404, real ships changed the experience of real Athenians radically and in specific ways, the ship is much more than a metaphor for the Athenian city. From its unanticipated naval victory at Salamis to its spectacular defeat at Syracuse – the city's fortunes rose and fell with its ships.

Similarly, walls are a key part of the Athenian civic experience, both lived and imagined. The Long Walls emerged from Athens' decision to embrace its navy and helped motivate the Athenians to develop a more urban civic environment, decorated with fine temples and shrines and yet cut off from its self-

64 Indeed, as many cultural critics (e.g., Foucault 1994 and Gilroy 1993) have shown, ships are enduring and productive conceptual images in a variety of cultural and historical contexts, useful for engaging with profound issues of civic transformation and change.

sufficient agricultural past. Walls protected Athenians from Spartan attacks, but they also condemned an overcrowded urban population to death from disease, and the ambiguity of walls in the Athenian experience explains why they function as such a useful conceptual tool as well.[65]

In addition, these texts show us that ships and walls are inherently connected and mutually implicated in the Athenian civic imagination. Herodotus' account of the wooden wall oracle frames fundamental questions about what truly defines the city as a debate over ships or walls as the best mode of defense for the city. The priest in *Oedipus the King* remarks that neither a city of ships nor one defined by walls is worth ruling if empty of men; but he does so, as does Herodotus, all the while keeping both models for the physical city in play. Thucydides' masterful account of the Sicilian expedition, on the other hand, exploits the ambiguities inherent in both ships and walls as models for the city rather than the choices between them. In his account, defensive walls turn into tools of aggressive siege warfare while his ship of state combines the infrastructural implications associated with the ship in the texts of Sophocles and Herodotus with their potential for mobility and destruction. Nicias' ship of state is "up in the air," simultaneously occupying a position in competing narratives of escape and imperial aggression, caught between the story of abandoning the old city and founding a new one.

Finally, what emerges from this focus on elements of civic infrastructure – on ships and walls – is a fascination with the possibilities and consequences of its failure or even absence. In all three texts, men emerge as an important third element in this discussion – either as inhabitants of or as metaphors for

[65] See Tracy 2000 for a collection of studies on the city wall cross-culturally. Cf. Goldhill 2007: 130: "The story of the Athenian citizen is constantly interwoven with the normative construction of the insider and the outsider, and the myth or ideology of the polis inevitably makes walls a charged space of definition." In a recent essay on the Achaean wall in the *Iliad*, Porter 2011 makes a similar observation about the doubleness of the wall; he argues that its poetic value lies in the way the wall is both made and obliterated.

the city's infrastructure. Their presence challenges or supplements the adequacy of physical infrastructure – Herodotus' ships of state are fully manned or occupied (πεπληρωμέναι); Thucydides' Nicias emphasizes the key role of men in defining the city, explaining that an empty city is no city at all, or certainly not, as Sophocles' priest says, one worth ruling. Perhaps even more striking, however, is their absence. The specter of desolation or depopulation (ἐρεμία) hovers powerfully at the edges of all of our texts. What is left when its ships and walls no longer exist?

This after all is the key question, and each of these texts draws upon the ambiguity of figurative language, the fusion of mythic and historic registers, and the complexity of overlapping narrative structures as part of what I have called a "poetics of infrastructure," in order to demolish and rebuild the city of Athens as a way to begin to find answers. The texts engage in a sustained discussion of what defines the city – ships, walls, or the men who inhabit them – a discussion that would be particularly pertinent to inhabitants of a city that had been physically destroyed not once but twice in less than a century. Xenophon tells us that at the end of the Peloponnesian war, while others wanted to raze the entire Athenian city, the Spartans offered to make peace on the condition that the Athenians destroyed their city walls and surrendered all but twelve of their ships – which they did, effectively erasing these iconic elements of the city's infrastructure.[66] And yet, through the texts of Herodotus, Sophocles, and Thucydides, we still have access to the civic imaginary of late fifth century Athens. These texts engage in the rich process of civic *poesis*, or reconstruction, and as long as they continue to be read and performed, the ships, walls, and men that inhabit them will continue to make and remake the Athenian city.

66 Xen. *Hell.* 2.2.23. And once the Athenians voted to accept these terms, Xenophon describes (2.2.24) how the Spartans "with great enthusiasm began to tear down the walls to the music of flute-girls, thinking that that day was the beginning of freedom for Greece."

Bibliography

Ahl, F. 1991. *Sophocles' Oedipus. Evidence and Self-Conviction*. Ithaca and London.

Allison, J. 1997. "Homeric Allusions at the Close of Thucydides' Sicilian Narrative." *AJP* 118:499–516.

Anderson, M. 1997. *The Fall of Troy in Early Greek Poetry and Art*. Oxford.

Avery, H. 1973. "Themes in Thucydides' Account of the Sicilian Expedition." *Hermes* 101.1:1–13.

Barker, E. 2006. "Paging the Oracles: Interpretation, Identity and Performance in Herodotus' History." *Greece and Rome* 53.1: 1–28.

Bassi, K. 2007. "Spatial Contingencies in Thucydides' History." *CA* 26.2: 171–217.

Binder, J. 1984. "Parthenon West Pediment: Poseidon." In *Studies Presented to Sterling Dow on his Eightieth Birthday*, ed. A. L. Boegegold and K. J. Rigsby. Durham, NC. 15–22.

Bollack, J. 1990. *L'Oedipe roi de Sophocle. Le texte et ses interpretations*. 4 vols. Lille.

Bowden, H. 2005. *Classical Athens and the Delphic Oracle*. Cambridge.

Broadhead, H. D. 1960. *Aeschylus: Persae*. Cambridge.

Bushnell, R. 1988. *Prophesying Tragedy. Sign and Voice in Sophocles' Theban Plays*. Ithaca and London.

Campbell, D. 1986. "Ship imagery in the *Oedipus Tyrannus*." In *Greek Tragedy and its Legacy*, ed. M. Cropp, E. Fantham, and S. E. Scully. Calgary. 115–20.

Conwell, D. 2008. *Connecting a City to the Sea. The History of the Athenian Long Walls*. Leiden.

Connor, W. R. 1984. *Thucydides*. Princeton.

Cresswell, T. 2004. *Place. A Short Introduction*. London.

Croally, N. T. 1994. *Euripidean Polemic: The Trojan Women and the Function of Tragedy*. Cambridge.

Dawe, R. D. 1982. *Sophocles: Oedipus Rex*. Cambridge.

de Certeau, M. 1984. *The Practice of Everyday Life*. Berkeley, Los Angeles, and London.

Dewald, C. 2005. *Thucydides' War Narrative. A Structural Study*. Berkeley, CA.

Dewald, C. and J. Marincola (eds.) 2006. *The Cambridge Companion to Herodotus*. Cambridge.

Dougherty, C. 1993. *The Poetics of Colonization: From City to Text in Archaic Greece*. New York.

Dover, K. J. 1965. *Thucydides. Book VI*. Oxford.

1965. *Thucydides. Book VII*. Oxford.

Earle, M. L. 1901. *The Oedipus Tyrannus of Sophocles*. New York.

Euben, P. 1986. "The Battle of Salamis and the Origins of Political Theory." *Political Theory* 14: 359–90.

Evans, J. A. S. 1982. "The Oracle of the 'Wooden Wall.'" *CJ* 78.1: 24–29.

Ferrari, G. 2000. "The Ilioupersis in Athens." *HSCP* 100: 119–50.

Fontenrose, J. 1978. *The Delphic Oracle: Its Responses and Operations with a Catalogue of Responses*. Berkeley and Los Angeles.

Fornara, C. 1971. "Evidence for the Date of Herodotus' Publication." *JHS* 91: 25–34.

Foucault, M. 1994. "Different Spaces." In *Essential Works of Foucault*, series ed. P. Rabinow. Vol. 2, *Aesthetics, Method, and Epistemology*, ed., J. Faubion. New York. 175–85.

Georges, P. B. 1986. "Saving Herodotus' Phenomena: The Oracles and the Events of 480 B.C." *ClasAnt* 5: 14–59.

Gilroy, P. 1993. *The Black Atlantic: Modernity and Double Consciousness*. London and New York.

Goldhill, S. 2007. "What's in a Wall?" In *Visualizing the Tragic. Drama, Myth, and Ritual in Greek Art and Literature. Essays in Honour of Froma Zeitlin*, ed. C. Kraus, S. Goldhill, H. P. Foley, and J. Elsner. Oxford. 127–47.

Gomme, A. W. A. Andrewes, and K.J. Dover. 1970. *An Historical Commentary on Thucydides*. Vol. IV. Oxford.

Greenwood, E. 2006. *Thucydides and the Shaping of History*. London.

Grene, D. 1987. *The History. Herodotus*. Chicago and London.

Griffith, D. 1993. "Oedipus Pharmakos? Alleged Scapegoating in Sophocles' *Oedipus the King*." *Phoenix* 47: 95–114.

Hanson, V. D. 1996. "Introduction." In *The Landmark Thucydides. A Comprehensive Guide to The Peloponnesian War*, ed. R. B. Strassler. New York. ix–xxiii.

Harrison, E. 1981. "Motifs of the City-Siege on the Shield of Athena Parthenos." *AJA* 85.3: 281–317.

Harrison, T. 2000. "Sicily in the Athenian Imagination: Thucydides and the Persian Wars." In *Sicily from Aeneas to Augustus: New Approaches in Archaeology and History*, ed. C. Smith and J. Serrati. Edinburgh. 84–96.

Harvey, D. 1996. *Justice, Nature and the Geography of Difference*. Cambridge, MA.

Hedrick, C. 1995. "Thucydides and the Beginnings of Archaeology." In *Methods in the Mediterranean: Historical and Archaeological Views on Texts and Archaeology*, ed. D. B. Small. Leiden. 45–88.

Holladay, A. J. 1987. "The Forethought of Themistocles." *JHS* 107: 192–97.

Hornblower, S. 1996. *Greek Historiography*. New York.

2008. *A Commentary on Thucydides*. Vol. III. Oxford and New York.

Hurwit, J. 2004. *The Acropolis in the Age of Pericles*. Cambridge.

Jebb, R. 1966. *The Oedipus Tyrannus of Sophocles*. Cambridge.

Jones, N. F. 2004. *Rural Athens under the Democracy*. Philadelphia.

Kamerbeek, J. C. 1967. *The Plays of Sophocles. Part IV. The Oedipus Tyrannus*. Leiden.

Knox, B. 1956. "The Date of the *Oedipus Tyrannos*." *AJP* 77.2: 133–47.

1957. *Oedipus at Thebes*. New Haven, CT.

Koster, W. J. W. 1975. *Scholia in Aristophanem*. Pars II. Fasc I: *Scholia Vetera et Recentiora in Aristophanis Vespas*. Gronigen.

Longo, O. 1974. "Ad Alceo 112, 10 LP: per la storia di un topos." *Bollettino dell'Instituto de filologia greca* 1:211–28.

1975. "La polis, le mura, le navi (Tucidide VII 77.7)." *Quaderni di storia* 1: 87–113.

Loraux, N. 1986. *The Invention of Athens. The Funeral Oration in the Classical City*. Cambridge, MA.

1993. *The Children of Athena: Athenian Ideas About Citizenship and the Division Between the Sexes*, trans. C. Levine. Princeton.

Macan, R. W. (1908) 1973. *Herodotus: The Seventh, Eighth, and Ninth Books*, 3 vols. London. (Repr. 1 vol.)

Maurizio, L. 1997. "Delphic Oracles as Oral Performances: Authenticity and Historical Evidence." *CA* 16.2: 308–34.

Mitchell-Boyask, R. 2008. *Plague and the Athenian Imagination*. Cambridge.

Morris, I. 1987. *Burial and Ancient Society: The Rise of the Greek City-State*. Cambridge.

Müller, C. 1984. *Zur Datierung des sophokleischen Ödipus* (No. 5. Akad. d. Wiss. ud Literatur). Mainz.

Ober, J. 1985. "Thucydides, Pericles, and the Strategy of Defense." In *The Craft of the Ancient Historian*, ed. J. W. Eadie and J. Ober. Lanham, MD, New York, and London. 171–88.

2001. "Thucydides Theoretikos/Thucydides Histor: Realist Theory and the Challenges of History." In *Democracy and War: A Comparative Study of the Korean War and the Peloponnesian War*, ed. D. R. McCann and B. S. Strauss. Armonk, NY and London. 273–306.

Page, D. 1955. *Sappho and Alcaeus; An Introduction to the Study of Ancient Lesbian Poetry*. Oxford.

Palagia, O. 1998. *The Pediments of the Parthenon*. New York.

Parke, H. W. and D. E. W. Wormell. 1956. *The Delphic Oracle*. 2 vols. Oxford.

Parry. A. 1969. "The Language of Thucydides' Description of the Plague." *BICS* 16: 106–118.

Pelling, C. 2000. *Literary Texts and the Greek Historian*. London and New York.
Robertson, N. 1987. "The True Meaning of the 'Wooden Wall.'" *CP* 82: 1–20.
Porter, J. 2011. "Making and Unmaking: The Achaean Wall and the Limits of Fictionality in Homeric Criticism." *TAPA* 141: 1–36.
Rader, R. 2009. "'And Whatever It Is, It Is You': The Autonomous Self in Aeschylus' *Seven Against Thebes*." *Arethusa* 42: 1–44.
Rengakos, A. and A. Tsakmakis (eds.). 2006. *Brill's Companion to Thucydides*. Leiden.
Rood, T. 1999. "Thucydides' Persian Wars." In *The Limits of Historiography*, ed. C. Kraus. Leiden. 141–68.
Rosenbloom, D. 1993. "Shouting 'Fire' in a Crowded Theater: Phrynichos' *Capture of Miletos* and the Politics of Fear in Early Attic Tragedy." *Philologus* 137: 159–96.
Rozario, K. 2005. "Making Progress Disaster Narratives and the Art of Optimism in Modern America." In *The Resilient City. How Modern Cities Recover from Disaster*, ed. L.J. Vale and T.J. Campanella. Oxford, 27–54.
Rusten, J. 2009. *Thucydides*. (Oxford Readings in Classical Studies). Oxford.
Shear, T. L. 1993. "The Persian Destruction of Athens." *Hesperia* 62.4: 383–482
Shephard, J. T. 1920. *The Oedipus Tyrannus of Sophocles*. Cambridge.
Silk, M. S. 1974. *Interaction in Poetic Imagery: With Special Reference to Early Greek Poetry*. New York and London.
Strauss, B. 2004. "Flames over Athens." *Arion* 12.1: 101–16.
Taylor, M. 2010. *Thucydides, Pericles, and the Idea of Athens in the Peloponnesian War*. New York.
Thomas, R. 2006. "Intellectual milieu of Herodotus." In *The Cambridge Companion to Herodotus*, ed. C. Dewald and J. Marincola. Cambridge. 60–75.
Thompson, H. A. 1981. "Athens faces Adversity." *Hesperia* 50.4: 343–55.
Thompson, N. 2001. *The Ship of State. Statecraft and Politics from Ancient Greece to Democratic America*. New Haven, CT and London.
Tracy, J. D. (ed.). 2000. *City Walls. The Urban Enceinte in Global Perspective*. Cambridge.
Vernant, J. -P. 1988. "Ambiguity and Reversal: On the Enigmatic Structure of Oedipus Rex." In *Myth and Tragedy in Ancient Greece*, J. -P. Vernant and P. Vidal-Naquet. New York. 113–40.
1991. "Speech and Mute Signs." In *Mortals and Immortals: Collected Essays*, ed. F. I. Zeitlin. Princeton. 303–17.

Vickers, M. 2008. *Sophocles and Alcibiades*. Ithaca.

Yeager, P. 2007. "Introduction: Dreaming of Infrastructure." *PMLA* 122.1: 9–26.

Zeitlin, F. I. 1990. "Thebes: Theater of Self and Society in Athenian Drama." In *Nothing to Do with Dionysos? Athenian Drama in Its Social Context*, ed. J. Winkler and F. I. Zeitlin. Princeton. 130–67.

5

CORINTH, COURTESANS, AND THE POLITICS OF PLACE

Kate Gilhuly

Bewitching Arts of the Courtesans

Corinth is called wealthy because of its commerce, since it is situated on the Isthmus and is master of two harbors, of which one leads straight to Asia and the other to Italy; and it makes easy the exchange of merchandise from both countries that are so far distant from each other... and also the duties on what by land was exported from the Peloponnesos and what was imported to it fell to those who held the keys. And to later times this remained ever so. But to the Corinthians of later times still greater advantages were added, for also the Isthmian games which were celebrated there used to draw crowds of people. And the Bacchiadae, a rich and numerous and illustrious family, became tyrants of Corinth, and held their empire for nearly two hundred years, and without disturbance reaped the fruits of commerce; and when Cypselus overthrew these, he himself became tyrant, and his house endured for three generations ... and the temple of Aphrodite was so rich that it owned more than a thousand temple slaves (*hierodouloi*), courtesans (*hetairai*) whom both men and women had dedicated to the goddess. And therefore it was also on account of these

> women that the city was crowded with people and grew rich; for instance, the ship captains freely squandered their money, and hence the proverb, "not for everyman is the voyage to Corinth." Moreover it is recorded that a certain courtesan said to the woman who reproached her with the charge that she did not like to work or touch wool: "Yet such as I am, in this short time I have taken down three webs."
>
> (Strabo 8.6.20–21, trans. Hamilton)

Strabo's dense description records fascinating commentary about Corinth, a major travel hub, where people could indulge in expensive pleasures. The emphasis he places on prostitution in his description would not surprise anyone familiar with Corinth's reputation in the ancient world, particularly Athens. Indeed, the Athenians associated Corinth so closely with prostitution that the verb comic poets derived from the city's name, *korinthiazesthai*, "to play the part of the Corinthian," means either to be a courtesan (*hetairein*) or, in the case of a male, to be a pimp (*mastropeuein*).[1]

While there is no doubt that a culture of prostitution flourished in ancient Corinth, the pervasive association of Corinth with courtesans in Athenian representations (and those, like Strabo's, that are influenced by them) suggests that the notion of Corinth as a place of prostitution was significant as a discourse as well as a reality. For Athens, like every other port city, had prostitutes of its own. While Strabo identifies the numerous courtesans at Corinth who helped make the city so wealthy as *hierodouloi hetairai*, temple slaves/courtesans, the only other reference to sacred prostitutes in Corinth is preserved in a fragment of Pindar recorded by Athenaios (13.573–4). Since the Athenian sources do not mention sacred prostitution at Corinth, the highly fraught and intriguing topic of sacred

1 The verb *korinthiazesthai* is found in six comic fragments (one by Aristophanes), and a noun related to it, *korinthiastes* (translated as "whoremonger") is the title given to two comic plays – one by Aristophanes and another by Poliochus.

prostitution – What was it? Did it really exist? – is not directly relevant to our discussion.[2]

Putting aside the reality of sex trade in ancient Corinth, this essay poses these questions: How does a place get a reputation? Why does Athens consistently talk about Corinth as a place of prostitution? What does this tell us about Athens? Benedict Anderson has revealed the degree to which communities are constituted by imagined relations that are constructed through narrative.[3] In this essay, I try to understand how one community, Athens, differentiated and defined itself through its imagination of an Other place, Corinth. I want to uncover what it was about Corinth that produced a people who, from an Athenian perspective, could be encapsulated by the idea of high-class prostitution, and further to think about how this image served the construction of Athenian civic identity.

My conception of place has been influenced by the work of cultural geographers like David Harvey who argues that place is a social construct.[4] For Harvey, a place can be the locus of the imaginary, a site for the negotiations of social relations, a form of discourse and of power, and at the same time a material practice.[5] This essay focuses on a discursive aspect of Corinth, a component of place that exists in the imaginary, and yet impacts the way a lived place is experienced. Tim Cresswell emphasizes the primacy of place to identity, suggesting that we think of place as a "'necessary social construction' – something we have to construct in order to be human."[6]

2 Budin 2008a and Beard and Henderson 1997 argue that sacred prostitution never existed in antiquity; Yamauchi 1973, Del Chiaro and Biers 1986: 20, Kurke 1996, and Davidson 2009: 360–62 argue that it did exist. While Budin's argument that the supposed eastern origins of this practice do not actually refer to sacred prostitution, is sound, neither her argument nor Beard and Henderson's is convincing regarding the testimony of Pindar and Strabo.

3 Anderson 2006: 8.

4 Harvey 1996: 293.

5 Harvey 1996: 294.

6 Creswell 2004: 33. Insofar as this chapter deals with the construction of place in the Athenian imaginary, it has similar concerns to Dougherty's contribution to this volume. Where Dougherty looks

If place and the social are crucially interdependent, it follows that gender and sexuality are implicated in the construction of place. What was it about Corinth that made one think of sex, sensuality, and paying for it? And why was this a good way to think about Corinth in Athens? An underlying assumption of this chapter is the notion that places are essentially relational: "place is a site of relations of one entity to another and it therefore contains 'the other' precisely because no entity can exist in isolation."[7] Significantly, the courtesan's identity is fundamentally that of a woman available for a relationship. For Athenians to project the image of Corinth as a land of courtesans is to conceive of Corinth in relation to themselves.

My argument begins with the Athenian comic discourse about Corinth, but as will become clear, the imaginary Athenian version of Corinth is related to its geography and culture, as well as the social and political climate of the place.[8] Yi-Fu Tuan has discussed the way that language has at times been overlooked in studies of place, noting that language has the power to make things visible that were once overlooked. "What once was a mere marker on the horizon can be transformed, by imaginative narration, into a vivid presence."[9] Indeed, Athenian language shaped ideas about Corinth and the way its history was written for centuries to come. Thus while my argument concentrates on courtesans, it could not be written without Corinth. It explores the intangible, the imaginary, the discursive aspect of the identity of a place, its reputation, its representation.

at Athens, self-construction, however, I am concerned with Athens' relationship to another place, and argue that the projection of alterity onto Corinth eventually subsumes its creator.

7 Harvey 1996: 261.

8 Like Worman's essay in this volume (Chapter 6), my discussion focuses on the construction of place on the comic stage. Where her argument considers dramatic space and its relationship to poetic discourse, I am ultimately interested in the way comic discourse penetrated the Athenian cultural imaginary more broadly conceived.

9 Tuan 1991: 689.

If we read the source material carefully, it becomes clear that representations of Corinthian courtesans tend to cluster around specific themes: tourism and hospitality; capitalizing on maritime commerce; issues of gender; working with the body; and oligarchy. Significantly, all of these themes relate to tensions in the Athenian polis, especially as it redefines itself over the course of the fifth century as a maritime empire, a city with a new port, a democracy defined in relation to oligarchy, an intellectual and cultural hub that receives multitudes of tourists for both ritual and secular tourism. In what follows, I proceed thematically, examining representations of Corinthian courtesans in conjunction with Athenian attitudes about Corinth and its culture, at the same time exploring how Athens constructs its own identity in opposition and identification with Corinth.

Servant to Strangers

Corinth was a passageway for numerous people and thus also numerous cultures. Aelius Aristides celebrates Corinth as a common city for all Greeks:

> It receives all cities and sends them off again and is a common refuge for all, like a kind of route or passage for all mankind, no matter where one would travel, and it is a common city for all Greeks, indeed, as it were, a kind of metropolis and mother in this respect. (Ael. Arist. *Orat.* 46.22)

Although Aristides wrote under the Roman Empire, the idea of Corinth as a place through which many people passed was by no means new. At least from the seventh century BCE onward, Corinth was host to many travelers and came to be known for its embrace of strangers.

And, in fact, many of these strangers would have been Athenians, since Corinth is not far from Athens, about fifty-five miles. The speech "Against Neaira," provides a sense of the ease of travel between Corinth and Athens in the ancient world:

the prosecution mentions a number of illustrious Athenians who consorted with Neaira while she lived in Corinth, including Xenokleides the Poet, and also notes that she accompanied Metaneira to Athens, when Lysias the Sophist sponsored her initiation into the Mysteries. Likewise, in Euripides' *Medea*, when Medea is plotting her revenge against Jason and her escape from Corinth, Aegeus, King of Athens, conveniently happens to be traveling through Corinth. His serendipitous appearance on stage has seemed puzzling to critics since Aristotle and a threat to the organic unity of the plot.[10] While the happenstance arrival of Aegeus conflicts with the conventions of tragedy, it conforms perfectly to the idea of Corinth as a travel hub.

Corinth's wealth is frequently ascribed to its ability to capitalize on its location, and to serve the needs of those passing through. An epigram written for Xenophon bills Corinth as *philoxenos*, or welcoming to foreigners, and in his epinician ode for Xenophon of Corinth, Pindar describes Corinth in the following way:

> τρισολυμπιονίκαν
> ἐπαινέων οἶκον ἥμερον ἀστοῖς
> ξένοισι δὲ θεράποντα, γνώσομαι
> τὰν ὀλβίαν Κόρινθον, Ἰσθμίου
> πρόθυρον Ποτειδᾶνος, ἀγλαόκουρον
>
> In praising a house with three victories at the Olympic games, I shall come to know wealthy Corinth, gentle to townsmen, while a servant to foreigners, door-way of Isthmian Poseidon, city of glorious children. (Pindar *O.* 13.1–5)

The image of Corinth that emerges from Pindar's poem is that of a rich city with a well-developed hospitality industry that is a gateway to sea travel. Pindar's choice of the words ξένοισι θεράπων is significant, for the willingness of the Corinthians to cater to those passing through the city (and to grow rich doing it) is key to the city's reputation. The connotation of

10 Arist. *Poet.* 1461b20.

subservience in this kind of business, which I have emphasized in my translation, also left its mark on the image of Corinth.

The ability of the Corinthians to make the most of their geographic location is often given sexual parameters: in Aristophanes' *Thesmophorizousai,* Cleisthenes likens the transport of ships across the isthmus to describe the antics of Euripides' kinsman as he struggles to appear like a woman: "That's some isthmus you've got there, buddy. You shuttle your cock back and forth more than the Corinthians." (*Thesmo.* 647–48.). This eroticizing of geography employs a logic similar to that which underlies the adage about Corinth that likens a sailing voyage to a tryst with a courtesan and may also relate to the treatment of the Corinthian envoy in Aristophanes *Lysistrata*. When she arrives on stage, the women admire her, remarking that she is attractive in the front and back (Ar. *Lys.* 76–91). Perhaps they are treating this unnamed woman in the same way as Lampito and the Boiotian before her, as embodying features and prejudices of the land she hails from. Lampito is vigorous and hale, since the Spartans are so strongly associated with military discipline. The Boiotian woman is admired for her beautiful plain, since Boiotia is so rural. The Corinthian is attractive on both sides because her front and backsides stand for the gulfs on either side of the isthmus.[11] Although nothing in the text overtly indicates it, a scholiast notes that the Corinthian is a *porne,* or prostitute. As Strabo mentions, the Isthmian games brought many tourists to the vicinity of Corinth, and this also takes on a sexual implication in the record of a courtesan named Isthmia.

The wealth of Corinth and its reputation as a host city are linked to its geographic location because it is the gateway to the sea. Corinth is a place of transition, a way station for people traveling, both by land and by sea. Of course, during the Classical period, Athens itself received many visitors, drawn to the city for all variety of reasons including trade, culture, and

[11] For an in depth discussion of the dynamics of this passage, see Gilhuly 2009: chapter 4.

ritual practices. Athenian humor about Corinth's hospitality culture must thus on some level be a projection of an evolving Athenian self-identity. Anxiety about tourism finds expression in an anecdote about Lais preserved by Athenaios. Diogenes the Cynic (412–323 BCE) questions Aristippus about his relationship to Corinth's most famous courtesan:

> "Ἀρίστιππε, κοινῇ συνοικεῖς πόρνῃ. ἢ κύνιζε οὖν, ὡς ἐγώ, ἢ πέπαυσο." – καὶ ὁ Ἀρίστιππος· "ἆρά γε μή τί σοι ἄτοπον δοκεῖ εἶναι, Διόγενες, οἰκίαν οἰκεῖν ἐν ᾗ πρότερον ᾤκησαν ἄλλοι;" "οὐ γάρ," ἔφη. "τί δὲ ναῦν ἐν ᾗ πολλοὶ πεπλεύκασιν;" "οὐδὲ τοῦτο," ἔφη, "οὕτως οὖν οὐδὲ γυναικὶ συνεῖναι ἄτοπόν ἐστιν ᾗ πολλοὶ κέχρηνται."

> "Aristippus, you are living with a common prostitute. Either be a Cynic then, as I am, or stop yourself." And Aristippus said, "It doesn't seem out of place to you, does it, to live in a house in which others have lived previously?" "No," he said. "What about a ship in which many others have sailed?" "Not that, either," he said. "In the same way then, it is not at all out of place to be with a woman whom many have enjoyed." (Ath. 13.588)

In his response to Diogenes, Aristippus compares Lais to a temporary residence and a passenger ship. Like a hotel, or a means of transportation, the prostitute exists for the traveler's short-term enjoyment. Diogenes' characterization of Lais as a common *porne* defines her as belonging to the realm of immediate and fleeting gratification, and this is at odds with the fact that Aristippus is living with her (συνοικεῖς). Aristippus recasts his cohabitation with Lais in the idiom of the traveler to justify his behavior to Diogenes, and in the process makes explicit the link between the courtesan and tourism – both accommodate transient and temporary enjoyment. Along with its reputation as a center of hospitality, an aura of promiscuity and degradation haunts the representation of Corinth – perhaps a tacit acknowledgment of the mystification that inheres in the kind of social relationships forged by the service industry, or anxiety about people who cater to visitors.

The Gender of Corinth

At a very basic level, the homology of Corinthians to *hetairai* effeminizes them. While Philetairos' and Poliochos' plays *Korinthiastes* derive their title from a male agent, often translated as whoremonger, the overwhelming usage of this stereotype refers to Corinthians as *hetairai* – that is, again, as high-class female courtesans. This gendering was clearly consolidated during the classical period. In a discussion of what would constitute the best physical regimen for the guardians, Plato enumerates pleasures that should be avoided, such as Attic pastry, the Syracusan table, and a Sicilian array of delicious tidbits. When he adds another pleasure to the list, Plato plays on the link between Corinth and the Greek word for maiden, kore: "You would censure the Corinthian maid (Κορινθίαν κόρην) as the girlfriend of men who were to keep themselves fit?" (*Republic* 404d). In Aristophanes' *Wealth*, a text that has more than a passing interest in Corinthian *hetairai*, there is a reference to Corinthian courtesans that is suggestive in terms of gender:

> καὶ τάς γ᾽ ἑταίρας φασὶ τὰς Κορινθίας,
> ὅταν μὲν αὐτάς τις πένης πειρῶν τύχῃ,
> οὐδὲ προσέχειν τὸν νοῦν, ἐὰν δὲ πλούσιος,
> τὸν πρωκτὸν αὐτὰς εὐθὺς ὡς τοῦτον τρέπειν.

> And they say that the Corinthian courtesans,
> Whenever a poor man happens to make trial of them,
> They don't pay him any mind, but if he is rich,
> They turn their asshole straight to him.
> (Aristoph. *Pl.* 148–52)

The word πρωκτός, which I have translated rather bluntly as "asshole," is curious here. As Henderson notes, it is a "pure" obscenity – there is no double-entendre here and thus such a word would be heard in a context that adhered only to the most relaxed standards of linguistic decorum. Further, this usage represents one of three times this word, common in the discussion of homosexuality, is applied to women. The other two have attenuating circumstances, one is in a comic compound (Peace 976) and the other (*Lys.* 1148), is used to describe a Spartan man's desire for Reconciliation, the mute nude in Aristophanes' *Lysistrata*. The Spartan ambassador's nation-state affiliation rendered him a consummate pederast (and thus obsessed with backsides), in much the same way Corinthians were thought of as prostitutes. According to Henderson, "πρωκτός was the *vox propria* for the anus in comedy and has an exclusively male (and therefore usually homosexual) reference ... Its low tone assured that even in the absence of a joke its mere mention could be counted on to raise a laugh."[12] Dover explains the seeming gender transposition of the *proktos* from male to female in a positivist way, claiming that it "suggests that *hetairai* may commonly have insisted on anal intercourse as a simple contraceptive measure."[13]

Rather than read a trace of ancient contraceptive practice in this passage, I would suggest that the reference to Corinthian courtesans is a joke. When they are assigned *proktoi*, the audience is cued to read them as male, and if all Corinthian *hetairai* have *proktoi*, then Aristophanes is drawing on the stereotype of all Corinthians as courtesans and the implication that they are willing to do anything for money. The veil of the stereotype is pulled away for a moment to reveal the joke's referent – that is, all Corinthians are subsumed under the label of *hetairai*.

Indeed *Wealth*, which is noted for its reticence about contemporary politics, is laced with references to Corinth, which alternate between allusions to Corinth's representation as a city of courtesans with the recent establishment there of a mercenary army. Wealth feeds the foreign troop at Corinth (173); Philonides, who is characterized as distinctly unattractive in comic anecdotes preserved by the scholiasts, is said to have won the love of the famous Corinthian courtesan Lais because of his money (179). Later Cario sings about Circe, who recently in Corinth turned Philonides' comrades into

12 Henderson 1991: 205.

13 Dover, 1989: 101.

wild boars (303–15). The scholiast informs us that Circe is Lais, and we are left to wonder why Cario is singing about Philonides' comrades instead of Philonides himself. Van Leeuwen's suggestion, that Philonides was associated with the mercenary army at Corinth helps to make sense of this. If Philonides' comrades are to be understood as mercenaries, then we can understand their transformation into boars, for in Aristophanes' writing, boars were symbolic of both lechery (*Pl.* 1024) and ferocity in battle (*Lys.* 1255). Stationed in Corinth, with their money earned through military service, both of the metaphorical meanings of the boar are relevant to these comrades – they could visit courtesans with the money they earned fighting. On this reading, Cario's song conflates Corinth's association with courtesans and the establishment there of a mercenary army. Intriguingly, the chorus winds up claiming that they will hang up Circe by the balls (312), in another gender conversion that suggests that the seductress of Corinth refers to a man.

The degree to which Corinth was identified by its female prostitutes is further attested in the sayings of Stratonikos, preserved in Athenaios. He refers to Herakleia, known for its male prostitutes as Ἀνδροκόρινθος, Man-Corinth, punning on Akro-Corinth, the high hill overlooking the isthmus. He goes on to say he was circumspect coming out of the gates of Herakleia because it was like exiting a *porneion*, or brothel (Ath. 8.351 c-d). The fact that Herakleia needs to be gendered masculine implies that the unmarked term, Corinth, is gendered feminine.

The effeminization of the Corinthians is fundamental to their association with courtesans. On the comic stage, where citizen-actors wore phalloi and performed for a male audience (at least putatively), this is clearly an expression of dominance. In a suggestive analysis of Eupolis' Cities, Ralph Rosen has argued that the feminine gender of the cities in that play allows Eupolis to embody the relationship between city and ally as one between a collection of males (the Athenians) and an allegorized female ally.

THE GENDER OF WORK

The characterization of all Corinthians as courtesans implicitly affirms the Athenian's phallic agency and their ability to dominate Corinth. But there is more to the courtesan than gender. For courtesans were women who worked. A common rhetorical formulation for the job of the courtesan is ἐργάζεσθαι τῷ σώματι, to work with one's body (e.g., [Dem.] 59.20). There is a spatial component to work, whether one works in or outdoors, and the location of work is gendered. Physical, outdoor labor was not value-neutral for the Athenians. Military service and farming were noble, manly pursuits, but craftsmanship and trade, things that could transpire indoors, were stigmatized. Corinth, however, fostered a culture that encouraged industry and thus capitalized on its position as a center of trade. A hint of the pejorative connotations that were associated with Corinthian socioeconomic attitudes can be found in a digression in Herodotus' description of the stratification of Egyptian society. After noting that the warrior class is forbidden from engaging in retail trade (*banausia* 2.165), he says,

> εἰ μὲν νῦν καὶ τοῦτο παρ᾽Αἰγυπτίων μεμαθήκασι οἱ Ἕλληνες, οὐκ ἔχω ἀτρεκέως κρῖναι, ὁρέων καὶ θρήικας καὶ Σκύθας καὶ Πέρσας καὶ Λυδοὺς καὶ σχεδὸν πάντας τοὺς βαρβάρους ἀποτιμοτέρους τῶν ἄλλων ἡγημένους πολιητέων τοὺς τὰς τέχνας μανθάνοντας καὶ τοὺς ἐκγόνους τούτων, τοὺς δὲ ἀπαλλαγμένους τῶν χειρωναξίέων γενναίους νομιζομένους εἶναι, καὶ μάλιστα τοὺς ἐς τὸν πόλεμον ἀνειμένους: μεμαθήκασι δ᾽ὦν τοῦτο πάντες οἱ Ἕλληνες καὶ μάλιστα Λακεδαιμόνιοι. ἥκιστα δὲ Κορίνθιοι ὄνονται τοὺς χειροτέχνας.

> I cannot say for certain whether the Greeks learnt this from Egypt, because I see that the Thracians, Persians, Scythians, Lydians and almost every non-Greek people also regard those who learn a trade and their descendants as the lowest stratum of society, as opposed to those who have nothing to do with artisanship and especially those who concentrate on warfare. However that may be, all the Greeks have adopted this attitude, with artisans coming

in for the most contempt in Lacedaemon, and the least in Corinth. (Hdt. 2.167, trans. Waterfield)

Here craftsmanship and trade are opposed to warfare, and the Corinthians are singled out as the most permissive of this implicitly unmasculine lifestyle. While Herodotus might not be considered a direct conduit for Athenian attitudes, it seems clear that Athenians and those concerned with Athenian policy constituted an important component of his audience. Herodotus' representation of Corinth was certainly known in Athens, and seems to tap into a shared set of beliefs. Indeed, the same denigration of craftsmen is found in a variety of other Athenian sources. Aristotle says that the best *polis* would not make a *banausos* (craftsman) a citizen (Pol. 1278a8).

Similarly, Plato in his Republic excludes craftsmen (*demiourgoi*) and farmers (*georgoi*) from participating in government. Xenophon depicts Socrates elaborating beliefs similar to those that inform Herodotus' opposition between artisanship and warfare:

> ἀλλὰ καλῶς, ἔφη, λέγεις, ὦ Κριτόβουλε. καὶ γὰρ αἵ γε βαναυσικαὶ καλούμεναι καὶ ἐπίρρητοί εἰσι καὶ εἰκότως μέντοι πάνυ ἀδοξοῦνται πρὸς τῶν πόλεων. καταλυμαίνονται γὰρ τὰ σώματα τῶν τε ἐργαζομένων καὶ τῶν ἐπιμελομένων, ἀναγκάζουσαι καθῆσθαι καὶ σκιατραφεῖσθαι, ἔνιαι δὲ καὶ πρὸς πῦρ ἡμερεύειν.τῶν δὲ σωμάτων θηλυνομένων καὶ αἱ ψυχαὶ πολὺ ἀρρωστότεραι γίγνονται. καὶ ἀσχολίας δὲ μάλιστα ἔχουσι καὶ φίλων καὶ πόλεως συνεπιμελεῖσθαι αἱ βαναυσικαὶ καλούμεναι. ὥστε οἱ τοιοῦτοι δοκοῦσι κακοὶ καὶ φίλοις χρῆσθαι καὶ ταῖς πατρίσιν ἀλεξητῆρες εἶναι. καὶ ἐν ἐνίαις μὲν τῶν πόλεων, μάλιστα δὲ ἐν ταῖς εὐπολέμοις δοκούσαις εἶναι, οὐδ᾽ἔξεστι τῶν πολιτῶν οὐδενὶ βαναυσικὰς τέχνας ἐργάζεσθαι.

> "Well said, Kritoboulus," he said, "for indeed the banausic arts, as they are called, are disparaged, and are, reasonably enough, held in complete disdain in our states. For they utterly ruin the bodies of the workmen and the foremen, forcing them to sit still and stay indoors, and in some cases to spend the day by the fire. The effeminization of their bodies accompanies a weakening of their

> minds. Moreover, these so-called banausic arts leave no leisure for attention to one's friends and city, so that their practitioners seem bad at dealing with friends and bad defenders of their country. In fact, in some of the states, and especially in those reputed to be versed in the arts of war, it is not even possible for any of the citizens to work at banausic arts." (Xen. *Oik.* 4.2–3)

Xenophon states explicitly that artisans' bodies are effeminized (θηλυνομένων), because their work brings them indoors and keeps them stationary. Their lack of leisure renders them incapable of pursuing social, political, and military affairs. But it was not only elite intellectuals like Plato and Xenophon (who suffered at the hands of the demos) who availed themselves of this stereotype; Aristophanes, whose comedies were produced for the entire citizen body, also associates craftsmen with women in the *Ekklesiazousai*. When a large group of cross-dressing women infiltrate the assembly, Chremes, a citizen, mistakes them for artisans, noticing how pale-faced the assembly had been the morning the women took over (Ar. *Ekkl.* 382–88). The association of pale skin with the feminine is the product of the gendering of space that Xenophon describes in the *Oikonomikos* – "for a woman it is more honorable to stay inside than to spend time in the field, whereas for a man it is more shameful to abide inside than to pursue things outdoors" (Xen. *Oik.* 7.30). Artisanship is gendered feminine because it demands time spent indoors and is opposed to the masculine pursuits of farming, politics, war, and social interaction. Put in these terms, craftsmanship is effeminate and opposed to democratic masculinity. I will take up the political dimension of this characterization in the next section.

The Corinthians were perceived as effeminate because they did not stigmatize artisanship, and thus many citizens were involved in trade. The pejorative associations that the Athenians made with craftsmanship – that it forced one to work inside and that it was effeminizing – conform to the image of the *hetaira*. Indeed one crucial element that separates the identity of the *hetaira* from the *porne* is space: the *porne* is a streetwalker, where often the courtesan's status is determined by the more rarefied, indoor context in which she operates.[14] Thus Socrates goes to visit the illustrious Athenian courtesan Theodote in a private home, attended by her mother and maids. As *hetairai,* the Corinthians are imagined as women who work with their bodies indoors. The emasculating quality of the Corinthian work culture is pointedly elicited at the end of the Strabo passage, cited earlier:

> καὶ δὲ καὶ μνημονεύεταί τις ἑταίρα πρὸς τὴν ὀνειδίζουσαν, ὅτι οὐ φιλεργὸς εἴη οὐδ᾽ἐρίων ἅπτοιτο, εἰπεῖν 'ἐγὼ μέντοι ἡ τοιαύτη τρεῖς ἤδη καθεῖλον ἱστοὺς ἐν βραχεῖ χρόνῳ τούτῳ.'
>
> Moreover it is recorded that a certain courtesan said to the woman who reproached her with the charge that she did not like to work or touch wool: "Yet such as I am, in this short time I have taken down three webs." (8.6.20 trans. Hamilton)

The joke here hinges on wordplay that does not come across in English. The courtesan's claim that she has taken down three webs can also mean that she has lowered three masts, or cavorted with three ship captains. Instead of weaving, woman's work par excellence, the courtesan is busily "lowering masts." Her indoor work even allows her to get the better of the real men doing manly work in Corinth, where the courtesan is on top and the men are emasculated.

NOT FOR EVERY MAN THE VOYAGE TO CORINTH

At the beginning of this essay, I quoted a famous proverb about Corinth: "Not for every man is the voyage to Corinth." As I have argued throughout this chapter, the courtesan is a conceptual medium through which the Athenians could talk about many topics; and in this case too I think there is more to this proverb than the idea that Corinth was a port town where it was easy to

14 Davidson 1998: 112–14.

squander money on the numerous and expensive courtesans. The very notion of the expensive courtesan is intriguing. For recent studies investigating depictions of prostitution have noted that a discursive opposition coalesced around the distinction between the *hetaira* and the *porne*:[15] the *hetaira* was associated with an elite discourse that embraced a culture of pleasure, cultivated eastern, specifically Lydian luxury, and was associated with gift exchange and undemocratic politics, while the *porne* was associated with the agora, democracy, and monetary exchange. The point of this opposition is to allow for the representation of contest between these ideological strata, and so we often see a woman represented as a *hetaira* only to be reconfigured and traduced in the image of a *porne*. Frequently, depictions of Corinthian courtesans blend elements from each pole of this discourse in a suggestive way. For they are generally identified as *hetairai* in order to denote elitism and sophistication, with the implication that they cultivate an eastern aura; yet at the same time, there is a ubiquitous emphasis on the mercenary ways of these women, so that any prestige communicated in the name of *hetaira* is immediately undermined by the persistent references to the high prices they charge. The mercenary *hetaira* is in a sense demystified, like an expensive gift with the price tag still on it.

As Strabo and Herodotus record, Corinth was ruled by oligarchs and tyrants: The Bacchiadai, Cypselus, and Periander. Since Herodotus' treatment of various Greek polis has been thoroughly discussed elsewhere, I only make brief reference to provide background for my argument. While Herodotus describes tyranny and other types of tyrannical rule in various Greek states, the treatment of Corinth diverges distinctly from the other mainland Greek polis, especially Sparta and Athens. Sparta never lived under a tyrant (5.92), although it did suffer under Cleomenes, who was a transgressive ruler par excellence.

15 Derived from the verb *pernemi*, which means to sell. For a discussion of prostitution terminology, see Dover 1989: 20–21; Davidson 1998: 109–26; Kurke 1999: 178–83; McClure 2003: 11–25; Gilhuly 2009: 13–14.

Nonetheless, the Spartans are resolutely free. Athens, on the other hand, was once ruled by tyrants – Peisistratus and his successors. Herodotus' narratives about the Peisistratids are constructed teleologically to explain how the Athenians freed themselves from tyranny. Although Corinth also only experienced a brief stint of tyrannical rule, Herodotus never depicts the Corinthians throwing off the yoke of the master. The enduring image of the Corinthians in the *Histories* is as a city that has more in common with the slavish barbarian, willingly submitting to despotic rule, than with the freedom-loving Greeks.

The costliness of the Corinthian courtesans suggests a contrast to a narrative that describes Solon's establishment of cheap prostitutes readily available for all Athenians as a democratic measure:

σὺ δ'εἰς ἅπαντας εὗρες ἀνθρώπους Σόλων
σὲ γὰρ λέγουσιν τοῦτ᾽ ἰδεῖν πρῶτον νόμον
δημοτικόν, ὦ Ζεῦ, πρᾶγμα καὶ σωτήριον,
(καί μοι λέγειν τοῦτ ἐστίν ἁρμοστόν, Σόλων)
μεστὴν ὁρῶντα τὴν πόλιν νεωτέρων
τούτους τ᾽ ἔχοντας τὴν ἀναγκαίαν φύσιν
ἁμαρτάνοντάς τ'εἰς ὅ μὴ προσῆκον ἦν,
στῆσαι πριάμενόν τοι γυναῖκας κατὰ τόπους
κοινὰς ἅπασι καὶ κατεσκευασμένας.
ἑστᾶσι γυμναί, μὴ 'ξαπατηθῇς· πάνθ᾽ ὅρα.
οὐκ εὖ σεαυτοῦ τυγχάνεις ἔχων, ἔχεις
⟨ἐρωτικῶς⟩ πως. ἡ θύρα'στ᾽ ἀνεῳγμένη.
εἷς ὀβόλος· εἰσπήδησον. οὐκ ἔστ'οὐδὲ εἷς
ἀκκισμός, οὐδὲ λῆρος, οὐδ᾽ ὑφήρπασεν,
ἀλλ'εὐθυς ἡν βούλει σὺ χὤν βούλει τρόπον.
ἐξῆλθες· οἰμώζειν λέγ᾽. ἀλλοτρία 'στι σοι.

But you, Solon, found a law for all men:
For they say that you were the first to think of
This democratic and saving matter
(And it is appropriate for me to say this, Solon).
When you saw that the city was full of young men,
And that they had a compulsive nature
To err against what wasn't theirs,

You established women in various places you'd bought
To be shared by all and ready for action.
They stand naked, lest you be fooled: look everything over.
You're not feeling well, you feel somehow turned on
—The door is open.
One obol. Leap in. There is no prudery, no foolishness,
she doesn't snatch herself away.
But immediately whom you want,
and the way you want her. You leave. Tell her to go to hell.
She's not yours.

(Philemon fr. 3)[16]

Philemon describes Solon's strategy for accommodating Athenian masculinity to an egalitarian social order. First, he identifies a problem: young Athenian men desire women that are not their own. If left to their own devices, these men will violate their fellow citizens' households through illicit sex. This system relies on a few crucial assumptions: that an Athenian man had the right to have sex; that he had the right to an intact household; and that women can be counted among a man's possessions. Solon's πρᾶγμα σωτήριον is to establish public prostitutes. Now young men can still have what does not belong to them, without impinging on the integrity of another man's household.

The prostitute retains a structural similarity to the original, prohibited pleasure of another man's woman: she has a place in the civic order, but has no relationship to the man who takes her. She is κοινὰς ἅπασι, common to everyone, but ἀλλοτρία'στι σοι not yours. The original conflict posed by the sexual double standard is absorbed in a paradox, the incorporated outsider. Solon's solution, making a commodity easily available and accessible to all Athenian men, is significantly described as a democratic (δημοτικόν) matter.[17]

16 This text is from Kassel-Austin, adopting Kaibel's suggestion to read νόμον in line two; ἔρωτικ]ς line 12 is the suggestion of Edmonds 1961.

17 While the most basic meaning of δημοτικῷν does not necessarily have the strong political charge conveyed by my translation

If we return to the semantic opposition between the *hetaira* and the *porne* and their association respectively with democracy and oligarchy, we can see that the restricted access to Corinthian courtesans suggested in the proverb by οὐ παντὸς, not for every man, might also be construed as an allusion to the Corinthian political constitution. For Corinth was never democratic and access to political affairs was never available to everyone, but only to the few. A prose version of the proverb was also found as an inscription on a red-figure vase painting in a private collection in Rome in 1847, which said: "Corinth is not for everyone" (οὐ παντὸς ἐστι Κόρινθος). This dates the adage to the fifth century BCE and more readily supports a political reading.[18] Horace also used this saying in a political context. He translated the maxim into Latin (*non cuivis homini contingit adire Corinthum, Epist.* 1.17.36) and embedded it in a poetic epistle urging his friend Scaeva to curry the favor of leading political men in order to further his public ambitions.

I suggest that there is a political meaning to Corinth's reputation, and that the maxim "Not for every man the trip to Corinth" casts a political comment into the language of travel and geography. The difficulty of gaining access to the courtesans also represents the idea that political participation in an oligarchy was the perquisite of the wealthy elite. Indeed, the exclusivity of Corinth's political elite was historically dramatized after the failure of the allied forces at Nemea in 394 BCE (Corinth, Argos, Athens, and Boeotia against Sparta) when the anti-Athenian contingent that had stayed behind at Corinth, that is the elite, closed the gates to the allied troops who were fighting for democratic ideals as they retreated back to Corinth in defeat (Xen. 4.2.23; Dem. 20.52–55). The cachet that might be conferred by the idea of exclusive access is undercut by the persistent references to the high

"democratic," the ascription of this invention to Solon suggests that it should be understood in this way. See Halperin 1990: 100.

18 Panofka 1847: 21–22.

cost of Corinthian *hetairai*. If the term hetaira communicates a sense of (here, political) exclusivity, it is undercut by the implication that for the right price Corinthian courtesans are eager to please. The mercenary hetaira encodes the image of Corinth as a place that is politically exclusive yet commercially promiscuous.

In its negation – "Not for every man" – the proverb about sailing to Corinth highlights Corinthian difference in its association with the courtesan. But as I said earlier, the courtesan is not only an other; she is also an internal Other,[19] and there is the possibility for (Athenian) male identification with this figure as well as difference. While the courtesan provides a way to highlight the political, cultural, and social differences that the Athenians perceived between their city and Corinth, there is also the suggestion of similarity and association in this figure. To put it another way, in order for Corinth to serve as a foil for Athens, similarities must exist between the two poleis. And indeed, Athens had so much in common with Corinth. Both were port cities that used their naval prowess to develop their wealth and influence in the Greek world; when Athenian democracy faltered, it was replaced by an oligarchy; both cities had similarly complex economies and were visited by many.

The Athenians developed the association of Corinth with prostitution just as the Piraeus was becoming a major urban center in Athens. Arguably it is this civic reconstruction that enabled Athens to supercede Corinth as a Greek political and economic power. Themistocles spurred the development of the harbor after the Persian invasion. It was completed in 477; under Pericles in 458–57, long walls were built that linked the city to the port, and so in the mid-fifth century, the Piraeus emerged as new focus of civic activity. The development of the Piraeus prompted a recalibration of the relationship of city to harbor. It is represented ambivalently and often with a degree of hostility in a number of texts in the fifth and fourth century,

19 Beard and Henderson 1997: 481.

frequently disparaging its commercial character:[20] Aeskhines notes that the Piraeus was the place where Timarchus sold himself as a prostitute to foreigners;[21] Aristophanes depicts it as home to prostitutes;[22] and Lysias mentions it as the center of operations for metic merchants who manipulate grain prices.[23] Jim Roy argues that the *astu* felt threatened by the distinctive culture that developed in the Peiraeus.[24] Sitta von Reden notes that the Piraeus was used in the "'rhetoric of 'otherness' – that is, the emphasis of the cultural limits of citizenship, which were essential for the self-consciousness of the Athenian *polis*."[25] Perhaps the Athenian discourse about Corinth as a place of prostitution is a projection of anxiety about the development of the Piraeus in the fifth century. For through the development of the Piraeus, Athens increased its nautical power, opened itself to more trade, welcomed more foreigners, and encouraged the growth of the hospitality industry that so often attends such civic development. As I have argued, these are the pejorative associations encoded in the figure of the courtesan, and they were as relevant to Athens as they were to Corinth. In the register of prostitution, Corinth served as a variegated site of affiliation and difference against which Athens could fashion its own identity.

THE COURTESAN HAS LEGS: THE AFTERLIFE OF LAIS

Given the background we have established, it is intriguing to consider the way the Athenian characterization of Corinth as a place of prostitution lived on in later sources. Athenaeus records

20 For an examination of the evolution of the discourse of the Piraeus' place in the city see Dougherty 2010; Garland 1987: 161.

21 Aeschines 1.40.

22 Aristophanes *Peace* 1.65.

23 Lysias 22.14.

24 Roy 1998: 199.

25 von Reden 1995: 32.

an anecdote about the rivalry between Lais the Corinthian, and Phryne, the most beautiful Athenian courtesan:

> διαζηλοτυπουμένη δέ ποτε ἡ Λαΐς τῇ Φρύνῃ πολὺν ἐραστῶν ἔσχηκεν ὅμιλον, οὐ διακρίνουσα πλούσιον ἢ πένητα οὐδ᾽ ὑβριστικῶς αὐτοῖς χρωμένη.

Competing with Phryne at one time, Lais had a whole crowd of lovers, and she did not distinguish between rich and poor, nor did she treat them arrogantly. (Ath. 13.588e)

At first glance, we might think that the decline of Lais indicates that the political associations between Corinth and Athens and Lais and Phryne have been reversed. Now Lais panders to a crowd, and makes no distinctions, while Phryne is in high demand. But if we consider this passage in the context in which it is preserved, I think it suggests another way of understanding the power and the meaning of the Athenian discourse of Corinth as a place of prostitution.

This passage invites us to map a competition between Corinth and Athens directly onto the discourse of prostitution and supports the argument of reading both cities together. The fact that the two women are in competition points to a certain degree of parity, yet clearly Phryne has an advantage. Phryne was known as a true beauty and was associated with the rich and famous; she was reputedly the model for Praxiteles' Knidian Aphrodite and Apelles' Aphrodite Rising from the Sea. Hypereides, a fourth-century orator, confesses to loving her more than even the most expensive courtesan Myrrhine, whom he kept in Athens. He defended her on a capital charge, and won his case upon exposing her breast in court (Ath 13.590). Phryne is a courtesan associated with the culture and politics of fourth-century Athenian democracy.

Lais was Corinth's most famous courtesan; her name was included in a list of *hetairai* by Anaxandrides in his *Gerontomania* (Ath 13.570d). Indeed there were two Laises of Corinth, one who lived during the Peloponnesian war, and one who lived later in the fourth century.[26] Understandably the two are confused, and often the anecdotes about "Lais" don't indicate which one is being referred to. For the purposes of this aspect of my argument, since I am considering the discourse of Corinthian courtesans as received in the Roman Empire, the distinction does not really matter.

The power and arrogance of Lais is captured in an epigram: ἡ σοβαρὸν γελάσα καθ᾽Ἑλλάδος. "She laughed haughtily at Hellas" (Plato *Epigr.* 15). The connection between Lais and Corinth was so strong that Eriphos, a contemporary of Antiphanes, a comic poet who wrote in the middle period, declares in his *Peltast*, when he praises the bounty of a generous table, "Such delights, o Syrian, neither Corinth nor Lais proffered, nor are they fare of the abundant Thessalian hosts, of which my hand has not been deprived" (Ath. 4.137d). According to W. S. Anderson, this quip implies "that Lais is equal to Corinth, the place of her practice, in wealth and in practical power."[27]

If we consider the passage where Athenaeus talks about the competition between Lais and Phryne in light of the other details his text records about the professional career of Lais, it suggests that the theme of the accessibility and overexposure of Lais may have had a different connotation during the Second Sophistic, and one that speaks to the power and attraction of the original Athenian discourse.

We first get a glimpse of Lais even before she became a courtesan: Apelles brought her to a symposium and his friends chided him for bringing a maid, a *parthenos*, not a *hetaira*. He retorted: "μὴ θαυμάσητε, εἶπεν, ἐγὼ γὰρ αὐτὴν μέλλουσαν εἰς ἀπόλαυσιν μετ᾽οὐδ᾽ ὅλην τριετίαν καλὴν δείξω." / "Don't be shocked, for I will show you that she will be, after not even a whole three year's time, a beautiful woman for the purpose of pleasure" (Ath 13.588d). But also included in the

26 On the dates of the two different Corinthian courtesans named Lais, see Anderson 1986: 45.

27 Anderson 1986: 47.

Deipnosophistae are passages concerning Lais that focus on her old age. Philetairos' *Huntress* is recalled in which she is reported to have died "fucked out" (βινουμένη, Ath. 13.587e), and her demise as depicted by Epicrates in his play, intriguingly titled *Anti-Lais*, is also preserved:

αὕτη γὰρ ὁπότ᾽ἦν μὲν νεοττὸς καὶ νέα,
ὑπὸ τῶν στατήρων ἦν ἀπηγριωμένη,
εἶδες δ᾽ἂν αὐτῆς Φαρνάβαζον θᾶττον ἄν·
ἐπεὶ δὲ δόλιχον τοῖς ἔτεσιν ἤδη τρέχει
τὰς ἁρμονίας τε διαχαλᾷ τοῦ σώματος,
ἰδεῖν μὲν αὐτὴν ῥᾷόν ἐστιν ἢ πτύσαι·
ἐξέρχεταί τε πανταχόσ᾽ἤδη πετομένη,
δέχεται δὲ καὶ στατῆρα καὶ τριώβολον,
προσίεται δε καὶ γέροντα καὶ νέον.

For when she was a chick and young,
made wild by the staters
You could get a glimpse of Pharnabazus more quickly than
of her
But since now she has run the course in years,
And the harmonies of her body relax,
It is easier to see her than to spit.
But now, she goes flying around everywhere
And takes both staters and triobols
And attaches herself to old and young alike. (Ath. 13.570 c-d)

As a young woman, Lais had the exotic allure of a Persian satrap, suggesting that she was powerful, luxuriant in an eastern style, and willing to grant audiences only sparingly. But as time wore on, she became ubiquitous, undiscriminating, and desperate for money. The sorry image of Lais depicted here, an aging courtesan, so far from the haughty beauty laughing at all of Greece, is a provocative symbol of loss and decay.

Considering the account of the competition between Phryne and Lais in this context, we can understand it in a different way: Lais, who was once sought-after and rare, grows old and is no longer desired by wealthy men. She has been replaced by Phryne, just as Corinth was replaced by Athens as a maritime power, tourist destination, and center for trade. Athens has been drawn into its own projection of Corinth.

Aulus Gellius records an anecdote about Lais that he ascribes to Sotion the Peripatetic, written in his *Horn of Amaltheia*:

> "Lais," inquit "Corinthia ob elegantiam venustatemque formae grandem pecuniam demerebat conventusque ad eam ditiorum hominum ex omni Graecia celebres erant, neque admittebatur nisi qui dabat quod poposcerat; poscebat autem illa nimuim quantum." Hinc ait natum ess illud frequens apud Graecos adagium:
>
> οὐ παντὸς ἀνδρὸς ἐς Κορινθον ἔσθ'ὁ πλοῦς,
>
> quod frustra iret Corinthun ad Laidem qui non quiret dare quod posceretur. "Ad hanc ille Demosthenes clanculum adit et ut sibi copiam sui faceret petit." At Lais μυρίας δραχμὰς poposcit – "hoc fecit nummi nostratis denarium decem milia." "Tali petulantia mulieris atque pecuniae magnitudine ictus expavidusque Demosthenes avertitur et discedens 'ego,' inquit, 'paenitere tanti non emo.'" Sed Graeca ipsa, quae fertur dixisse, lepidiora sunt: οὐκ ὠνοῦμαι, inquit, μυρίων δραχμῶν μεταμέλειαν.
>
> Lais of Corinth, he says, used to earn a lot of money because of the elegance and charm of her beauty, and was frequently visited by wealthy men from all over Greece; but no one was received who did not give what she demanded, and her demands were extravagant enough. He says that this was the origin of the proverb common among the Greeks:
>
> "Not every man may fare to Corinth town,"
>
> For in vain would any man go to Corinth to visit Lais who could not pay her price. "The great Demosthenes approached her secretly and asked for her favors, but Lais demanded ten thousand drachmas" – a sum equivalent in our money to ten thousand denarii. "Amazed and shocked at the woman's great impudence and the vast sum demanded, Demosthenes turned away, remarking as he left her: "I will not buy regret at such a price." But the Greek words which he is said to have used are neater; he

said οὐκ ὠνοῦμαι μυρίων δραχμῶν μεταμέλειαν. (Aul. Gell. 1.8.3, trans. J. C. Rolfe)[28]

This narrative suggests a near homology between Lais and Corinth, and links her to the proverb about Corinth. In this context, sailing to Corinth is explicitly linked to sleeping with Lais, a courtesan emboldened to demand an obscenely high price. The exchange with Demosthenes is telling too, for here we have Demosthenes, the consummate Athenian orator, long remembered and venerated for fighting for democracy and freedom against the imperialist Philip of Macedon, engaged with Lais, the courtesan-icon of Corinth. Demosthenes is seduced by what Corinth/Lais has to offer, but when he finds out her exorbitant fee, he quips "I will not buy regret for ten thousand drachmas." Lais' excessive demand propels Demosthenes to the mindset of the morning after. Foregoing the allure of an easy present pleasure he gives voice to its necessary consequence – remorse for what has been spent, nostalgia for the past. The failed negotiation between Demosthenes and Lais demonstrates that the courtesan is a figure of regret and nostalgia, representing squandered potential and loss.

In the classical period, Athens took the place that Corinth had previously occupied as one of the preeminent cities in Greece with the greatest maritime power. The image of Corinthians as courtesans encodes the Athenian perception that Corinth was an Orientalizing city of men willing to work with their bodies indoors, more luxuriant than warlike, excessively committed to pleasure, incapable of democracy, and if in any way like Athens, Corinth was a thing of the past, the subject of nostalgia. Although Athens constructed this version of Corinth to explore dangerous possibilities about its own identity, eventually Athens itself was drawn into the discourse about Corinth and prostitution. During the Roman Empire, we see

28 This narrative almost certainly merges the identity of the two Laises because the Greek adage, as I have suggested, originated in the fifth century, or shortly thereafter, and Demosthenes lived a generation younger.

that literary collectors adopted the Athenian discourse about Corinth as a way to think about both cities. These authors confirm my contention that Athens was constructing its own identity through the branding of Corinth, and also that this discourse reflects an anxiety about an internal other, the kind of place Athens was becoming as it grew into its identity as a port city. In the Roman context, the courtesan is used to represent the rise and fall of Corinth *and* Athens, the seductive allure of power and its inevitable demise.

Bibliography

Anderson, W. S. 1986. "Corinth and Comedy." Del Chiaro, M. and W. Biers eds. *Corinthiaca: Studies in Honor of Darrell A. Amyx*. Columbia, MO: 44–49.

Ahl, F. 1991. *Sophocles' Oedipus: Evidence and Self-Conviction*. Ithaca.

Balme, M. 1984. "Attitudes to Work and Leisure in Ancient Greece." *Greece and Rome* 31: 2.

Beard, M. and Henderson, J. 1997. "With This Body I Thee Worship: Sacred Prostitution in Antiquity." *Gender & History* 9.3: 480–503.

Buchan, M. 2008. "Too Difficult for A Single Man to Understand." *Helios* 35: 3–28.

Budin, S. 2008a. *The Myth of Sacred Prostitution in Antiquity*. New York.

2008b. "Simonides' Corinthian Epigram." *CP* 103: 335–53.

Cresswell, T. 2004. *Place. A Short Introduction*. London.

Davidson, J. 1998. *Courtesans & Fishcakes: the Consuming Passions of Classical Athens*. New York.

2006. "Making a Spectacle of Her(self): the Greek Courtesan and the Art of the Present." Feldman, M. and B. Gordon. *The Courtesan's Arts: Cross-cultural Perspectives*. New York: 29–51.

Davidson J. 2009. "Eros and Sexuality." In *Companion to Ancient History*, ed. A. Erskine. West Sussex. 352–67.

Del Chiaro, M. and W. Biers (eds.). 1986. *Corinthiaca: Studies in Honor of Darrell A. Amyx*. Columbia, MO.

Dillon, M. 1987. "Topicality in Aristophanes' PLOUTOS." *CA* 6.2: 155–83.

Dougherty, C. 2009. "Just Visiting: The Mobile World of Classical Athens." In *The Oxford Handbook of Hellenic Studies*, ed. G. Boys-Stones, B. Graziosi, and P. Vasunia. Oxford. 391–400.

2010. "Athens' Tale of Two Cities: Themistocles, Theseus, and the Construction of 'Place' in Fifth-Century Athens." In *When*

Worlds Elide. Classics, Politics, Culture, ed. K. Bassi and J. P. Euben. Lanham, MD. 137–60.

Dover, K. J. 1989. *Greek Homosexuality*. Cambridge.

Dunkle, J. R. 1969. "The Aegeus Episode and the Theme of Euripides *Medea*," *TAPA* 100: 97–107.

Edmonds, J. M. 1961. *The Fragments of Attic Comedy after Meinecke, Bergk, and Kock*, vol. 3a. Leiden.

Faraone, C. A. 2004. "Lysistrata the Priestess or the Courtesan? Female Stereotypes in Aristophanes' *Lysistrata*." In Faraone, C. A. and L. McClure. *Prostitutes and Courtesans in the Ancient World*. London and Madison, WI. 274–94.

Faraone, C. A. and L. McClure. (2006). *Prostitutes and Courtesans in the Ancient World*. London and Madison, WI.

Feldman, M. and B. Gordon (2006) *The Courtesan's Arts: Cross-cultural Perspectives*. New York.

Garland, R. 1987. *The Piraeus*. London.

Gilhuly, K. 2004. "The Phallic Lesbian: Philosophy, Comedy and Social Inversion in Lucian's *Dialogues of the Courtesans*." In Faraone, C. A. and L. McClure. ed. *Prostitutes and Courtesans in the Ancient World*. London and Madison, WI..

2009. *The Feminine Matrix of Sex and Gender in Classical Athens*. Cambridge.

Gould, J. 1989. *Herodotus*. New York.

Gray, H. 1996. "Herodotus and Images of Tyranny: The Tyrants of Corinth." *AJP* 117.3: 361–89.

Halperin, D. 1990. *One Hundred Years of Homosexuality and Other Essays on Greek Love*. New York.

Hartog, F. 1988. *The Mirror of Herodotus: the Representation of the Other in the Writing of History*. Trans. J. Lloyd. Berkeley and Los Angeles.

Harvey, D. 1996. *Justice, Nature and the Geography of Difference*. Cambridge, MA.

Henderson, J. 1991. *The Maculate Muse*. Oxford.

Hornblower, S. 2006. "Herodotus' Influence in Antiquity." In *The Cambridge Companion to Herodotus*, ed. C. Dewald and J. Marincola. Cambridge: 306–18.

Janko, R. 2007. "Pity the Poor Traveller: A New Comic Trimeter (Aristophanes?)." *CQ* 57: 296–97.

Kassel, R. and C. Austin. 1983. *Poetae Comici Graeci (PCG)*. Berlin and New York.

Kovacs, D. 1993. "Zeus in Euripides' *Medea*." *AJP* 114: 45–70.

Kurke, L. 1996. "Pindar and the Prostitutes, or Reading Ancient 'Pornography'." *Arion* (3rd series) 4.2: 49–75.

1999. *Coins, Bodies, Games, and Gold: the Politics of Meaning in Archaic Greece*. Princeton.

McClure, L. 2003. *Courtesans at Table: Gender and Greek Literary Culture in Athenaeus*. New York.

Murray, O. 1990. *Sympotica: a Symposium on the Symposion*. Oxford.

Panofka, T. 1847. Archäologisches aus Italien: Bericht an die Archäologische Gesellschaft zu Berlin. *Archäologische Zeitung* 5/6 = N.F. 1, Beil 2: 17–28.

Rosen, R. 1998. "The Gendered Polis in Eupolis' Cities." In *The City as Comedy: Society and Representation in Athenian Drama*, ed. G. W. Dobrov. Chapel Hill, NC. 149–76.

Roy, J. 1998. "The Threat from the Piraeus." In *Kosmos: Essays in Order Conflict and Community in Classical Athens*, ed. P. Cartledge, P. Millett, and S. von Reden. Cambridge. 191–202.

Salmon, J. 1984. *Wealthy Corinth: A History of the City to 338 B.C.* Oxford.

Sourvinou–Inwood, C. 1991. *Reading Greek Culture. Texts and Images, Rituals and Myths*. Oxford.

Stroup, S. 2003. "Designing Women: Aristophanes' *Lysistrata* and the Hetairization of the Greek Wife." *Arethusa* 37.1: 37–73.

Thomas, R. 2006. "The Intellectual Milieu of Herodotus." In *The Cambridge Companion to Herodotus*, ed. J. Marincola and C. Dewald. Cambridge. 60–75.

Tuan, Y. 1991. "Language and the Making of Place: A Narrative – Description Approach." *Annals of the Association of American Geographers* 81.4: 684–96.

von Reden, S. 1995. "The Piraeus – A World Apart." *Greece and Rome* 42.1.

West, S. 1999. "Sophocles *Antigone* and Herodotus Book Three." In *Sophocles Revisited*, ed. J. Griffin. Oxford. 109–36.

Williams, C. 1986. "Corinth and the Cult of Aphrodite." In Del Chiaro, M. and W. Biers eds. *Corinthiaca: Studies in Honor of Darrell A. Amyx*. Columbia, MO: 12–24.

Yamauchi, E. M. 1973. "Cultic Prostitution: A Case Study in Cultural Diffusion." In *Orient and Occident: Essays Presented to Cyrus H. Gordon*, ed. H. Hoffner. Neukirchen-Vluyn, Germany. 213–22.

6

MAPPING LITERARY STYLES IN ARISTOPHANES' *FROGS*

Nancy Worman

Early on in the action of Aristophanes' *Frogs*, when Dionysus asks Heracles to show him the way to Hades, he envisions the journey as one that follows a particular path through literal spaces (cf. ὁδούς, 113, 117). Heracles, in contrast, treats the path to Hades as a metaphor, and offers Dionysus a number of ways to commit suicide. Dionysus plays along, rejecting various modes as physically uncomfortable and confirming that he needs a quick route, not being much of a walker (ὡς ὄντος γε μὴ βαδιστικοῦ, 128; cf. 134). Lazy critic though he is, he finally requests the path that Heracles himself took, and the hero tells the god of drama in Athens precisely how to go in "real" (i.e., mimetic) space, though he warns him that it is a "massive passage" (ὁ πλοῦς πολύς, 136).[1]

Dionysus must, Heracles says, take the road (cf. 135) that goes through a big and bottomless marsh (λίμνην μεγάλην. . .ἄβυσσον) (137–38), a path that the Frog chorus that greets him deems the precinct of "Dionysus in the Marshes" (Διόνυσον. . .ἐν λίμναισιν, 216–17). He successfully crosses the *limnai*, accompanied by the Frogs' jocular reference to the "hangover crowd"

[1] By "mimetic" I mean space as it is demarcated onstage, which is "real" or actual (i.e., not metaphorical) within the fictional frame. Stanford *ad loc.* captures the onomatopoeic humor of the phrase ὁ πλοῦς πολύς as "a whale of a sail."

invading the precinct during the Anthesteria celebration (cf. ὁ κραιπαλόκωμος/ τοῖς ἱεροῖσι Χύτροις χω-/ ρεῖ κατ' ἐμὸν τέμενος λαῶν ὄχλος, 218–19). After this crossing, he arrives at flowering meadows for processional dances by new choruses of male and female celebrants that appear to be engaging in Mysteries "play" (παίζειν, 376, 392, 443, 452). Finally, he knocks on the door of "Hades," which eventually opens out onto the famous contest between the tragic poets Euripides and Aeschylus.

In *Frogs*, as elsewhere in Attic comedy, the multilayered semiotics of theatrical space and situation complicates the revelation of where one is, as well as with whom and to what end. On the dramatic stage quite generally any referential or deictic gesture points up a here and now that is first and foremost fictional, mimetic of some other time and place, and secondarily resonant with the actual space of the theater – say, on the south slope of the Acropolis, in the ritual swing of the springtime.[2] As Nick Lowe has explored, comic space, particularly in the hands of Aristophanes, often shapes a wild ride, from earth to the upper air, up and down the Acropolis, in and out of the city.[3] When, for instance, at the beginning of Aristophanes' *Birds*, Euelpides and Peisetaerus leave the city to seek out an avian Elsewhere, they trace within the confines of the Theater of Dionysus a non-urban, non-Athenian setting that is demarcated, at least initially, by thickets and birdsong. These two Athenians explicitly indicate their civic identities by means of this search for a not-Athens, a quiet space apart from the busy (and busybody) work of the city (*Av*. 40–45). Later in the play they will build their city in midair, having identified the vault of the sky (*polos*) as the proper place for their *polis* (179–84).

Such orchestrations of place and space engage a mapping of schemes, within topographies that are usually quite well known, that are imbued with intellectual and political

2 On theatrical referencing and semiotics, see Ubersfeld 1977; Elam 1980; Issacharoff 1989. See also Dougherty (Chapter 4 in this volume) for the ways in which "infrastructure" metaphors in tragedy and historical narratives reinforce Athenian civic identity.

3 Lowe 2006.

significance, and that constitute a unique intermingling of the aesthetic, the ethical, and the spatial. As with Chris Eckerman's exploration of Pindar's poetics of place and Tim Rood's demonstration of Xenophon's mastery of terrain (Chapters 1 and 2 in this volume), my discussion treats both landscapes and "landscapes." That is, I consider both the concrete stamp of culture on actual spaces and their shaping within dramatic literature and by onstage semiotics as resonant, value-laden settings. Though primarily fictional and only secondarily directly referential, these settings are not abstract spaces in any true sense; rather, they are deeply rooted in the ground of civic practices, organized and assigned value in relation to recognizable transactions in familiar places.

While *Birds* charts a path out of the city that dramatizes a political and intellectual terrain of fantastical proportions, *Frogs* traces the route to literary judgment as passing through a landscape that is grounded in local topographies and rituals of aesthetic and religious import. It thereby serves a unique function in explorations of space and place in Greek drama. At the outset of *Frogs*, the audience is offered the inside joke of "paths" on the comic stage – namely, that there are paths and "paths," including those that since Homer indicate a poetic "way" or style. In addition, within the semiotics of the dramatic stage, there is the path through mimetic space, which maps the "Athenian landscape" within the Theater of Dionysus; there are also the actual ritual spaces that the dramatic mimesis appears to reference, which accrete symbolic significance through this mapping. And then, again, there are paths that signal "ways" in some more inclusive sense – in this play especially those that intertwine the aesthetic, religious, and civic realms.[4]

4 Raymond Williams 1973, in his famous discussion of English pastoral poetry, emphasizes repeatedly the imbeddedness of the literary discourse in concrete social and economic relations to the fields, hills, roads, and towns that shape a community's everyday life. See also Buxton 1994; Schama 1995; and Cosgrove and Daniels (1988) 2002 on the imprint of culture on topography.

The action of the play first conflates modes of leave-taking ("getting to Hades") with a literal path from Heracles' house, across a marsh, and into the countryside. Scholars have usually highlighted the ritual significance of the play's changing spaces, and many regard its signposts as pointing importantly to features central to the Eleusinian Mysteries.[5] My aim in this discussion is not to dispute such connections, as these may well be vital to understanding the ritual shaping of Dionysus' journey to the Underworld. But *Frogs* is the only extant Attic comedy that features Athenian rituals to this extent, which in itself encourages questions about how the cultic references relate both to the localities mimed onstage and to the play's primary focus, literary judgment. I thus pursue here in tandem the literary, ritual, and topographical signposts that together shape the path that Aristophanes charts from city to countryside. Focusing on the first half of the play, I look to the stylistic conventions and rituals that were practiced locally in dramatic and fertility celebrations, some of which appear to have taken place in riverside spaces outside the city walls, such as along the Cephisus and the Ilissos. The latter becomes the more significant space for the literary critical tradition, and it cannot be happenstance that the topographical references in *Frogs* indicate that Aristophanes associates a burgeoning literary critical discourse with this riverside setting. In fact, his dramatization of this stylistic landscape was quite influential, since Plato and later literary critics situate their own stylistic discussions in relation to this same terrain.[6]

5 Segal's 1961 important article, despite its focus on Eleusis, does emphasize that Aristophanes also intertwines local rituals such as the Anthesteria and Lesser Dionysia; cf. Moorton 1989; Bowie 1993; Lada-Richards 1999; Edmonds 2004. As earlier scholars such as Tierney 1934 have emphasized, wartime impediments to the Eleusinian procession make it unlikely that Aristophanes would directly lampoon this prominent ritual, the loss of which must have been sharply felt by his Athenian audience. Note as well that the play never directly references Eleusis or the Eleusinian Mysteries.

6 Most evident in Plato's *Phaedrus*, Cicero's *De Oratore*, and Dionysius of Halicarnassus' *On the Style of Demosthenes*. Cf. Gilhuly's discussion of how places get reputations (Chapter 5 in this volume).

Why these discursive references should be located here has to do with a clustering effect in which ritual practices, topographical features, and poetic traditions converge to shape a fitting path for the comic enactment of literary judgment. As I show, along this route from city to (quasi-) countryside, choral songs that invoke local rituals separate out stylistically one from another, and these ritually inflected stylistic strands illuminate the significance of different spots on the way, prefiguring later distinctions between the tragic poets in relation to the urban-rural divide, to old and new modes, and to gender allegiances.

In the sections that follow, I first locate Aristophanes' scheme within the larger literary critical tradition that makes stylistic meaning out of rural and quasi-rural landscapes. Next, focusing in on *Frogs*, I highlight the features along the path out of the city that shape differences between choral modes and the poets' styles. I then consider the intersection of ritual practices and topographical details from the prospect of the Theater of Dionysus, the mimetic spaces of which resonate with the actual terrain that lies below the south slope of the Acropolis. Finally, I consider a specific set of aesthetic coordinates that grounds Aristophanes' intervention in this literary critical tradition: the styles that rivers can shape.

The Literary Critical Landscape

The literary critical journey that Aristophanes traces in *Frogs* takes its most prominent cue from poetic and ritual traditions that situate inspiration and fertile proliferation outside of the city walls, down along a valley stream or up a mountainside.[7]

[7] See, e.g., Harriott 1969; Connor 1988. Cf. Rosenmeyer 1969: 267–70 on the importance of the pastoral setting to ancient poetics. Discrete studies also treat (e.g.) path/road metaphors (Becker 1937; Lefkowitz 1963), rivers and springs in Alexandrian and Augustan poetry (Zetzel 1983), Hesiod's imagery (e.g., Pucci 1977), metaphors in Pindar (Steiner 1986), metaphors in Callimachus (Asper 1997), river imagery in Roman literature (Jones 2005), Greek landscapes (Snell 1955: chapter 16; Parry 1957; Elliger 1975), and the topos of

In ancient poetic and critical discourses more generally, metaphors that indicate distinctions among styles of writing often trace their lineaments by means of elements in rural settings.[8] Not only paths and roads, but also rivers, springs, and meadows in flower mark out stylistic differences in ancient literary usage. The shapes and spaces sketched in this way sometimes contrast rural and urban settings; and they are usually inhabited by those who can harvest, harness, or properly utilize the stylistic "resources," whether these are the lovely flowers of a meadowland or garden, the flowing waters of a river or mountain spring, or the life-sustaining wheat of a farmer's field.[9] Further, many of these formulations invoke what we might call a "landscape with figures" model: they delineate quasi-pastoral or rural spaces whose significant features generate metaphors (i.e., "figures") that serve inhabitants (i.e., "figures") who respond to, make pointed use of, and/or in some way have affinities with these features.[10]

Thus if, for instance, Pindar's poet culls flowers from the Graces' garden as an indication of his own metaphorical style (*O.* 9.28–29), so do the artistry and distinctive styles of Euripides and Aeschylus in *Frogs* correlate to various spaces along the path out of the city. Further, *Frogs* fashions a discriminating route for Dionysus by means of one of the oldest and most prominent metaphors in ancient literary conception – one

the *locus amoenus* (Curtius 1953: chapter 10; Schönbeck 1962; Hass 1998). Again, cf. Eckerman (Chapter 1 in this volume) on Pindar's landscapes.

8 Scholars have argued about "pastoral" versus "bucolic" as generic labels (e.g., Halperin 1983; more recently Fantuzzi 2006), but my focus more generally (Worman forthcoming) encompasses an inclusive discursive space, which runs from epic to prose and from mountain springs to city limits. Thus gardens (which may be essentially suburban), shepherds' paths (which at least mark human husbandry), and farmers' fields (which are cultivated spaces), though they may not all signify in the same ways, would all fall under this general interpretive distinction of non-urban space that is conceived of in tradition as natural/rural but is also marked by human use.

9 See, e.g., Hes. *Th.* 1–4, 29–31, Sapph. fr. 2.5–7, Pi. *P.* 6.1–4, Pl. *Phdr.* 237c4–238d3, Cic. *Orat.* 228–29, DH *Dem.* 32.10–15.

10 See further in Worman forthcoming.

that seems near dead in certain contexts, only to become fully invigorated in others. This path (ὁδός, οἶμος, κέλευθος) of words, a usage extending from Homeric epic to Roman rhetoric, is usually deployed as a means of indicating aesthetic and/or moral direction.[11] The Homeric poet refers to the epic song as a "path" (οἴμης, *Od.* 8.74; cf. 481, 22.347); in the *Homeric Hymn to Hermes* Apollo similarly calls the poetry that the Muses inspire "the shining path of song" (ἀγλαὸς οἶμος ἀοιδῆς, 451).[12] Hesiod, in some contrast, invokes the path of virtue that leads straight to justice (ὁδὸς ἑτέρηφι παρελθεῖν κρείσσων ἐς τὰ δίκαια, *WD* 216–17); later he warns that it is long, steep, and rough at the beginning (μακρὸς δὲ καὶ ὄρθιος οἶμος ἐς αὐτὴν/ καὶ τρηχὺς τὸ πρῶτον, 291–92). This kind of ethical landscaping initiates an adjacent tradition of tropes that mark the intersection of aesthetic and morally instructive tracks. Both the "path of song" image and that of the "virtuous path" are familiar from Pindar (e.g., τιμῶντες ἀρετάς/ ἐς φανερὰν ὁδὸν ἔρχονται, *Ol.* 6.72–73; cf. *Pyth.* 3.103, *Nem.* 7.51).[13] That is, path images tend to imply a complex and varying conjunction of superior aesthetics (i.e., the path of the poet's song, cf. ἐπίκουρον εὑρῶν ὁδὸν λόγων, *Ol.* 1.11) and moral direction.

In Aristophanes' comedies, the use of such idioms as the "path of words" (ὁδὸς λόγων) may parallel a sense of place (on stage, in life).[14] Markus Asper has argued that Aristophanes'

11 See Asper 1997: 23–26 for arguments and bibliography.

12 Although scholars have disagreed as to whether in Homeric epos these terms indicate a very strongly envisioned "way" for a particular poetic mode or story, the vocabulary of beginnings and movement certainly points in this direction. See Becker 1937: 5 vs. Snell 1955: 320–32 and Harriott 1969: 65; also Asper's discussion (1997: 23–26). Snell is largely interested in the *Scheiderwege* as an aporetic image (following Panofsky 1930).

13 This is the merest sketch, of course; for details see esp. Becker 1937: 50–100, who demonstrates the breadth and complexity of the Pindaric imagery; also Steiner 1986: 76–86; Asper 1997: 39–46; and Harriott 1969: 63, who compares Bacchyl. 5.176–78, 10.51–52, 19.1–8.

14 This is a dominant image in *Birds*, where the action begins with a pondering over which path will lead to a better polity underpinned by a "finer" ethical sense.

use of the *Wegmetapher* most often takes aim at the "new music," that is, the dithyrambic poets who indulge in overly modulated, elaborate cadences.[15] Such decadent aesthetics render the poetry like ants' tracks: minutely articulated and thus difficult to discern, as Euripides' In-law complains of Agathon's effeminate trilling in *Thesmophoriazusae* (μύρμηκος ἀτραπούς, Ar. *Thesm.* 100; cf. frs. 31 and 155 PCG).[16] Asper's arguments encourage us to recognize that Dionysus' very search for the path to Hades may be itself born of a decadent desire for the newfangled styles, embodied in *Frogs* especially by Euripides. That said, other Aristophanic characters make mocking reference to the "path of argument/song" (*Eq.* 1015, *Pax* 733, *Av.* 1374), and the dramatic action often foregrounds as well "roads/ways" (ὁδοί) as a means of "getting somewhere" (e.g., in argument, in plotting, in utopian endeavors).[17]

In *Frogs*, variants of ὁδός roughly match occurrences in other plays, but here the "path" emphasized repeatedly at the outset of the play ultimately marks out a "way" of literary judgment.[18] While other plays frequently engage in parodic critique of poets and their poetry (esp. Euripides in *Acharnians*, *Clouds*, *Thesmophoriazusae*), only *Frogs* explicitly proceeds toward and then enacts a contest between dramatic poets that the chorus enjoins in terms that reiterate the action: "Step onto the contested path!" (ἔπιτε δαΐαν ὁδόν, 897). This may in fact be the sole extant moment in Attic comedy where the path of choice is clearly that of literary critique rather than poetic mode, even as it wryly echoes earlier poets' intersections of style and moral

15 Asper 1997: 39–46.

16 But cf. *Clouds* 75–76, where the ὁδοῦ ἀτραπόν denotes merely a clever plan.

17 This is especially prevalent in *Knights* (demagogic plotting, 72, 253, 291, 621), and *Birds* (utopian planning, 4–29 *passim*, 994, 1004). Streets (ὁδοί) are also public stages, where citizens disport themselves in various embarrassing or manly postures (e.g., swaggering, clowning: *Knights* 348; *Clouds* 362, 964; *Wasps* 542).

18 N.b. Aristophanes does not appear to use other terms from archaic poetry for paths and roads (e.g., *keleuthos* appears only once: *Thesm.* 1100).

compass. The irony reverberates, since the play targets poetic judgment itself as a fraught and debased practice, placing the contest alongside other rituals related to fertility that appear to represent a turn toward older sensibilities and to offer a rural corrective.

Thus the route out of the city also charts an implicit politics that at one level sets an earlier, more rustic, and conservative perspective on civic virtues off against a latter-day, urbane, and dissolute one.[19] In this scheme the city, to put it in reductive but clarifying terms, is a corrupting space of dissembling, overly clever argument, while the country remains a reminder and source of simpler, more manly pursuits (e.g., farming, fighting). Such contrasts initially appear to mock the polished Euripides in contrast to the rugged old Aeschylus, but those that frame the action of *Frogs* extend beyond the familiar one between city and country. Indeed, these contrasts ultimately give way to innovations on the older poetic tendency mentioned above, which associates styles of composing and writing and those who might actively judge them with significant rural settings. The hybrid figure that emerges in the course of *Frogs*, the poet-critic who may lay claim to a novel mode, accretes stylistic features from various stops along the way and thereby transcends the urban-rural divide.

Frogs is not alone among Aristophanes' plays in offering up the journey out of the city as a necessary distancing that reveals contrasts among city players and that thus would give the traveler (were he suitably aware) his ethical and aesthetic bearings. But what of Dionysus? If pastoral spaces traditionally afford inspiration and increased artistic authority, the countryside does not seem to serve him very well, because he remains a buffoonish figure throughout the play.[20] As the god

19 See, e.g., McGlew 2002.

20 As both Stanford and Dover have noted: Stanford 1983: xxix–xxx; Dover 1993a: 39, 42–43. This is against those who see some growth in Dionysus, as a result of ritual (esp. Segal 1961: 209–11) or more general to see character development (e.g., Silk 2000). See also Bowie 1993: 228–53; Lada-Richards 1999; and Edmonds' 2004: 113–17 correctives. Further, this "initiation of Dionysus," while it may map

of both dramatic poetry and fertility and thus a central figure in poetic and life-cycle rituals (including the Lenaea, the Anthesteria, and the local Mysteries), Dionysus should be a fitting pivot between these realms and thus singularly likely to make much of his rustic surroundings. Similarly, the rituals that structure so much of his passage out into the country and down to Hades should underscore the importance of the pastoral or rustic settings for the poetic contests. In aesthetic terms, Dionysus is the art's internal measure, inhabiting and in some ways coincident with the rural spaces shaped by the play. He is also the one on the move, literally and figuratively; and so he should be able to realize the long view needed for effective critical judgment. And yet, the play instead offers up a loosely interconnected clutch of comic adventures that do little to inform the god's aesthetic understanding, and neither of the tragic poets survives the contest unscathed. Because comedy thrives on repeatedly upending established attitudes, and because comic protagonists are so mischievous, no clear path of judgment emerges. Dionysus chooses Aeschylus for a silly reason but one central to comedy, as the poet's profundity is reduced to a "weight of words," while the play's own critical perspective has more in common with Euripidean refinements. The god of tragedy, though playfully established as judge and tastemaker from the outset of the action, makes a poor master of critical idioms; in the end it seems to be comic enactment itself that achieves the critique.

THE MARSH AND THE MEADOWLAND

If literary critical discourse does originate at least in part in Hesiodic and Pindaric judgments that map the path of words and identify sites of inspiration as rural, Aristophanes' foregrounding of pastoral metonymies would be quite traditional

stages of the Mysteries on a "rites of passage" model, does not account very well for how the journey and the contest fit together. See Moorton 1989 for an attempt to read the *Frogs* as a drama that illustrates van Gennep's ([1909] 1981) stages of initiation.

in a general sense. His turn on the critical topography, on the other hand, is extremely subtle and innovative, especially in the vocabulary that it foregrounds to differentiate stylistic categories. The scene with which I began is a good example of the ritual shading he gives the route, while the one that it follows on sets up some novel stylistic parameters. The play opens with an en route critique of typical comedies as full of clichéd physical humor, bandied as Dionysus and Xanthias are approaching Heracles' house (βαδίζω, 23); but when Dionysus reveals his desire (πόθος, 53) for Euripides, the comic and critical target shifts to tragedy. This sparks some aesthetic elbowing, in which Heracles scoffs at Dionysus' trashy tastes when the god of dramatic poetry explains with the peevishness of a frustrated reader (cf. ἀναγιγνώσκοντί μοι/ τὴν Ανδρομέδαν, 52–54) his aim to bring Euripides back. Heracles questions why Dionysus is not satisfied with poets who are still alive, and Dionysus responds with a set of significant distinctions that illuminate his preferences, predominant among which are that Euripides is roguishly up for anything (πανοῦργος, 80) and fertile (γόνιμος, 96).[21] While both Heracles and Dionysus associate Euripides with chatter and wordiness (cf. λαλίστερα, 91; στωμύλματα, 92, cf. 930ff.) – labels that Aristophanes uses to lampoon styles he finds effeminized and decadent – Dionysus regards this feature as evidence of stylistic boldness.[22] Without Euripides around, he says, there is no one who "might bark out a noble phrase" (ῥῆμα γενναῖον λάκοι, 97). For Dionysus this is a gloss on *gonimos*, as is his designation "whoever utters this sort of daring thing" (ὅστις φθέγεται/ τοιουτονί

21 I.e., like a comic protagonist, perhaps. On the γόνιμος ποιητής, see Woodbury 1988: 181–82, who argues that this new coinage denotes not merely creativity but especially novelty; also Padilla 1992: 365–67. On *panourgos* sophistic types, see Worman 1999; at *Ran.* 1015 Aeschylus associates "marketplace wranglers" (ἀγοραίους).

22 This vocabulary is typically used to categorize Euripides and his sort (e.g., Socrates, *Frogs* 1492; cf. 826–27, 841, 917, 943, 954, 1071). See O'Sullivan 1992: 131–33; Worman 2008: 96–110. Lossau 1987: 239–41 emphasizes the contrast as one of "light" vs. "heavy" (*leptos* vs. *barus*).

τι παρακινδυνευμένον, 98–99), which he follows with some choice figures of speech.[23] Heracles mocks these, calling them "tricks" (κόβαλα, 104) and "utterly base" (παμπόνηρα, 106), but Dionysus is adamant in his support.

Thus before Heracles tells Dionysus where to go, he effectively tells him where to get off, opposing his decadent taste for Euripides with a tough guy's commitment to more traditional styles. But countering this staunch rejection is the enthusiast's love of novelty: Dionysus asserts a newfangled profile for his favorite poet, which combines boldness, invention, and a "fertile" mode. As careful readers of these terms have noticed, this combination is a hybrid and unusual one; for instance, a phrase like ῥῆμα γενναῖον sounds more suited to the elder poet Aeschylus' elevated modes, just as a label like *gonimos* would seem a fitting label for his fulsome style.[24] But Aristophanes appears to be redefining terms, and thus offering the possibility of a new type of daring, fertile style, one that he and Euripides share.[25] This mini-debate, the first of the literary critical contests in the play, anticipates important distinctions not only in the agon in Hades but also in Dionysus' iambic competition with the Frogs and sing-along with the Initiates. As a debate on the fly, en route to other bucolic and Mysteries-infused settings, it serves to initiate the action as literary critical and thereby shape the comic plot as essentially forged by such judgments.[26]

Insofar as it makes certain key stylistic claims, the debate also initiates questions as to where the fertile and roguish

23 This notion of a style being daring or dangerous is influential; cf. Arist. *Rhet.* 1405b5–11; Demetr. *De eloc.* 80; DH *Dem.* 2.28–31; Ps.-Long. 32.3.

24 See, e.g., Lossau 1987; Lada-Richards 1999: 242–47.

25 See further below and in the final section. Woodbury 1988 recognizes the redefinition, but draws different conclusions from it.

26 The pursuit of a poet, particularly Euripides, for help with the plot is a very typical Aristophanic ploy; cf. Dicaeopolis' pursuit of Euripides (or rather his props) in *Acharnians*, Trygaeus' appropriation of a Euripidean plot at the opening of *Peace*, and Euripides' own recourse to Agathon in *Thesmophoriazusae*.

Euripides belongs in this passage out of the city, past the Frog chorus' suburban pastorale, and through the flowering meadows of the Initiates, singing their anthems. And what about old Aeschylus, for whom Dionysus has not a thought? Or to put it another way, what does the marshy river setting have to offer stylistically, as opposed to the meadowland in flower?

Euripides and the Marshes

When Dionysus crosses the marsh, the Frog chorus parades its pastoral, Ur-poetic status. It is beloved of the Muses and fosters the reeds that make the instruments of Pan and Apollo:

> Ἐμὲ γὰρ ἔστερξαν εὔλυροί τε Μοῦσαι
> καὶ κεροβάτας Πάν, ὁ καλαμόφθογγα παίζων·
> προσεπιτέρπεται δ᾽ ὁ φορμικτὰς Ἀπόλλων,
> ἕνεκα δόνακος, ὃν ὑπολύριον
> ἔνυδρον ἐν λίμναις τρέφω.
>
> For the well-lyred Muses love me
> and horn-footed Pan, the player on the reed;
> And Apollo with his phorminx delights still more,
> for the sake of the reed, which watery
> lyre-bridge I nourish in the marshes. (229–34)

The setting of the chorus' song and the content of its claims are both very wet and patently bucolic. The "watery lyre bridge" (ὑπολύριον/ ἔνυδρον, 233), which is echoed in the second strophe by the "watery dance" in the swampy depths (ἔνυδρον ἐν βυθῷ χορείαν, 247), frames this style as one among the range suggested by rivers and streams. This one is light, bubbly, and playful rather than full flowing – that is, it is in keeping with both Aristophanic and Euripidean styles.[27] And yet it is also a "country" mode: the reeds go to make the shepherd's pipes of the woodland Pan and Apollo's phorminx, the instrument

27 See Biles 2002; Ruffell 2002.

that Hermes made from a tortoise that he found outside the door of his mountain home (cf. δόνακας καλάμοιο, *HHerm.* 47; φόρμιγγα, 64). What could this have to do with Euripides, who is represented later in *Frogs* as quintessentially urban? We might notice first that the song that the Frogs sing is metrically reminiscent of Euripides' preference for simple iambs and trochees, with a high degree of resolution and some "lekythia" endings.[28] Second, both of the Frogs' strophes contain echoes of Euripidean choral odes, and one depicts the Frogs' style as overly modulated like that of Euripides and other purveyors of the "new music."[29]

I want to consider this cluster of effects in more detail. It is an odd mix, with its little sing-song endings for which Aeschylus will later tease Euripides, as finishing off his iambs with domestic ditties: "a little sack, little bottle, little fleece" (καὶ κῳδάριον καὶ ληκύθιον καὶ θυλάκιον/ ἐν τοῖς ἰαμβείοισιν, 1203–04). Some scholars have seen a sexual joke here, one more suggestion, in this play full of them, that Euripides' style is lacking in manly force.[30] But this meter also shapes a marshy poetics that one might be inclined to call *gonimos*, because its tutelary deities are the riverside Muses,[31] a rustic Apollo, and the horned Pan, playing on his reed pipes.

The figures that consolidate the Frogs' claim to poetic authority feature also in some Euripidean choral odes that associate a sparkling pastoral style with young women's rites of passage. This locating of a particular song and dance type is especially vibrant in odes from Euripides' *Ion* (492–505) and *Iphigenia in*

28 See Parker 1997: 465–66.

29 Despite Segal 1961: 22, who thinks that the compounds suggest an affinity with Aeschylus. Cf. Radermacher 1954: 171–73.

30 So Dover *ad loc.*, citing ληκᾶν; cf. O'Sullivan 1992: 110, 150 on the complications of the imagery (*lekythos* sound may be deep and resonant elsewhere, e.g., Call. fr. 215 Pf.), but this does not at all fit the scene, despite O'Sullivan's careful relating of it to the "bombast" of tragedy (110).

31 Pausanias deems the Muses celebrated by the river "daughters of Ilissos"; see further below in the third section.

Tauris (1125–31) that describe ritual spaces on slopes of the Acropolis. The *Ion* offers the fullest scene:

ὦ Πανὸς θακήματα καὶ
παραυλίζουσα πέτρα
μυχώδεσι Μακραῖς,
ἵνα χοροὺς στείβουσι ποδοῖν
Ἀγλαύρου κόραι τρίγονοι
στάδια χλοερὰ πρὸ Παλλάδος
ναῶν συρίγγων
ὑπ' αἰόλας ἰαχᾶς
{ὕμνων} ὅτ' ἀναλίοις
συρίζεις, ὦ Πάν,
τοῖσι σοῖς ἐν ἄντροις,
ἵνα τεκοῦσά τις
παρθένος μελέα βρέφος
Φοίβωι....

O seats of Pan and the rock
abutting the cave-pierced Long Cliffs
where the three daughters of Aglauros
trod the dances with their feet
in the green courses before the temple
of Athena, when you, O Pan,
with the dazzling shouts
of the syrinxes' [hymns]
pipe in your caves,
where a certain maiden
unhappily bore
an infant to Apollo.... (492–505)

Here we have Pan piping, the daughters of Aglauros dancing and singing on green courses (στάδια χλοερά), and caves associated with both Pan and Apollo. The topography described is that of the area below the Erechtheion on the north slope, where there were caves dedicated to Pan and Apollo and a sanctuary of Aglauros (cf. *Ion* 936–38), as well as of Aphrodite.[32]

32 Wycherley 1978: 176–79. On the significance of the Acropolis and Aphrodite, see further below in section 3. Cf. E. *Hel.* 179–90, where Pan's rape of a nymph by the waterfall serves as an analogy

Compare also a strophe from *Iphigenia in Tauris*, which reiterates the central theme and players:

> συρίζων θ' ὁ κηρόδετος
> Πανὸς οὐρείου κάλαμος
> κώπαις ἐπιθωΰξει,
> ὁ Φοῖβός θ' ὁ μάντις ἔχων
> κέλαδον ἑπτατόνου λύρας
> ἀείδων ἄξει λιπαρὰν
> εὖ σ' Ἀθηναίων ἐπὶ γᾶν.

> Piping, the wax-bound reed
> of the mountain Pan
> will shout a beat for the oars,
> and Apollo the seer, holding
> the din of the seven-toned lyre,
> singing will lead you [Iphigenia] happily
> to the shining land of the Athenians. (1125–31)

Again the featured figure is the woodland, piping Pan (with his wax-bound reed) and now Apollo enters more fully into the song with his echoing seven-toned lyre; both are deployed as the Athenian welcoming committee – as facilitating, like the Frogs, passage to and from Athens. Such figures conventionally function as metonyms for older rustic modes that have been domesticated (i.e., brought inside the city) for Athenian ritual use.[33] While the actual terrain is not the same, both the Euripidean odes and the Frogs' song herald these rural deities as crucial to a vibrant transitional mode. The Frogs themselves

for Helen's snatching by Hermes. The *Bacchae*, the play with which many scholars think *Frogs* is intricately engaged, also contains choral imagery tracing outside erotic spaces, especially 403–16, which invokes Aphrodite, the Erotes, the Charites, and Pothos.

33 Cf. Wilson 1999 on the *aulos*. Moorton 1989: 314 considers this echo of the *IT* ode part of Aristophanes' overall parody of Euripidean style; he emphasizes, however, that the Frogs accompany the opposite passage (i.e., away from Athens). Note also that Plato's Ilissos setting in the *Phaedrus* includes Pan, the Nymphs, and the Muses; see Rosenmeyer 1962. Also *Bacch.* 951–52. On Pan in Athens, see Borgeaud (1979) 1988: 140–41, 151–56; on his ties to Demeter, 141–50.

are more fully hybrid: they may lay claim to Pan's pipe, but their own style is a heady mix of new and old, of inside and outside the city walls.

The second strophe of their ode, which again combines trochees, iambs, and lekythia, furthers the impression of a "pastoral," light, and lively guise: their song is one of "many diving melodies" (πολυκολύμβοισι μέλεσιν), "sparkling dance" (χορείαν/ αἰόλαν), – recall the "sparkling" (ἀιόλας) shouts of hymns in the *Ion* ode quoted above – and multi-bubblings (πομφολυγοπαφλάσμασιν) (244–49).[34] All of these elements point to a vivacious, even girlish, and overly modulated style, one that sounds like those featured in the "new music" of the dithyrambic poets Timotheus and Kinesias, as well as Euripides' later lyrics.[35] Critics of this new style (including Aristophanes) depict it as decadent and effeminizing; by activating images suggestive of pastoral erotics these bubbly melodies use old lyric metonyms to trace a new, fluid, and animated mode.[36]

If it is accurate to regard the Frogs as singing a song that is at least somewhat "Euripidean," this would mean that the poet's style is in certain crucial aspects mediated by pastoral, or rather "pastoral" – that is, the poetic mode that in Athenian ritual signals erotic adventure, passages from inside to outside, and the incorporation of outside as "outside" within the

34 Cf. the "Platonic" epigram, in which the frog is termed a "rain-loving, watery poet, servant of the Nymphs" (Τὸν Νυμφῶν θεράποντα φιλόμβριον ὑγρὸν ἀοιδόν, *Epigr*. 22 Page). On girlish play in erotic lyric poetry, see Rosenmeyer 2004.

35 For connections to Euripides and the "new music" see Defradas 1969; Campbell 1984; Moorton 1989: 313–14; also Csapo 2004 for more general observations and Power 2010 for the associations with cithara music. Many of the metaphors that Aristophanes and others use to describe (and usually lampoon) this style involve the vocabulary of bending or twisting (i.e., cognates of *kamptein* and *strephein*); later in *Frogs* "twisting" is a technique that Euripides teaches (στρέφειν ἐρᾶν, 957; cf. *Frogs* 775, *Clouds* 331–34, *Thesm.* 53–62 and further in Franklin 2013).

36 Note that Sappho is invoked later in the literary critical tradition as representative of this style (Demetr. *de Eloc*. 128–32, 167, DH *de Comp. Verb*. 23.1, 22.35).

city and/or within civic ritual.[37] I say this despite the fact that Aristophanes clearly figures Euripides as an urban type, in *Frogs* and elsewhere, and while recognizing that Aristophanes' critical discourse (with which Euripides is strongly associated) clearly has an urban and distinctly Athenian source. But Euripides himself serves as a mediating figure between city and country, just as this urban and urbane critical discourse makes use of pastoral details to show, with full comic irony, why Aeschylus' older, staunchly rustic, and "weightier" style might be valued more than the subtle, discerning style that Euripides shares with the critical mode itself. Aristophanes' association of Euripides with this space in between is both innovative and perspicuous, since if this is the space of criticism, Euripides is its perfect denizen, embracing as he does not only the rationalizing, refined mentality emphasized later in the play (cf. *Ran.* 954–58; cf. 1101–07) but also the feminized, agile, and "fertile" aesthetics that constitute this space's more subtle offerings. And yet given his alignment with this critical edge, Euripides cannot ultimately be Dionysus' choice for the poet to save the city, since the play also mocks the critical process as well as this judge's abilities to manage it to any effective end.

Aeschylus and the Meadowland

While Euripides' style may share features with marshy and feminine modes, and thus seem a liminal, newfangled creature (i.e., a literary-critical poet), Aeschylus clearly represents an older, grander, and more natural source of creativity. As Segal notes, the chorus and his interlocutors associate him with the Initiates and Mysteries ritual, and he follows suit. The Initiates' parodos, during which Dionysus travels with them through

[37] I use the terms "pastoral" and "bucolic" provocatively, since they usually mark a tradition that scholars take to start with Hellenistic poets and especially Theocritus, but the genre clearly shares features with erotic lyric, tragic choral odes (esp. Euripides'), and fourth-century epigram.

mimetic space, foreshadows Aeschylus' later invocations of Demeter and the Mysteries, as well as his use of elevated, hymnic modes and support of silent, pious entrances.[38] If we can detect a ranking between these two stops along the way, the Frogs' marshy setting may foster the reeds for the instruments of (pastoral) poetry, but the Initiates occupy the flowering meadows between the stream of "Acheron" and the door to "Hades," a paradise of a patently bucolic sort. The Initiates invoke the meadows repeatedly (344, 374, 449, 374; cf. 351[39]), sing a hymn to Demeter, and do so in an ionic meter with a rare feature (anaclomenon) used by Aeschylus.[40] Thus while the chorus may not retain their identity as initiates throughout the play, they do serve at this point in the play as decorous conductors of the god of poetry through the pastoral meadows and groves to Hades. The chorus calls upon Iacchos (the local Mysteries' ritual name for Dionysus) as the one who can complete the "long road" (πολλὴν ὁδόν, 402) of their ritual song and dance, a characterization that for some scholars must point to Eleusis. But, as we know from Hesiod, the "long road" may also suggest an elevated mode, both stylistically and morally, that would match with the more formal aspects of Aeschylus' grand style. By the end of the parodos the Initiates have entered meadows full of flowering roses where they "play" (παίζοντες) in the noblest choral style (τρόπον/ τὸν καλλιχορώτατον) (448–53). They also declare that they have sustained a pious manner

38 See Segal 1961: 218; Lada-Richards 1999: 247–54.

39 I.e., they seem to be moving through a number of flowery meadows in the course of the parodos. Dover compares the flowering meadows of Pindar's "paradise" (*O.* 2.70–75. fr. 29.3–5), but note that the hymnic strophes are answered by those full of comic play (*skôptein*, *paizein* used repeatedly, e.g., 374, 375, 393, 414, 417, 443, 452, vocabulary that infiltrates the "elevated" strophes by 440ff.). Thus language demarcates the space more as one of ritual procession (from marsh to meadow, meadow to grove, etc.), complete with play and *aischrologia*, important in fertility rituals involving Dionysus and Demeter (including the Eleusinian Mysteries). See Foley 1994: 45–46.

40 Parker 1997: 468.

(εὐ-/ σεβῆ...τρόπον, 456–58), the repetition of the word linking poetic and religious practice (as well as song and dance styles) in a lofty mode.

The second half of the play confirms the sense that it is Aeschylus who belongs in the older, more traditionally celebrated pastoral setting. Despite the scheme that marks out the path of literary judgment and situates both poets' styles in quasi-rural spaces, the distinctions that the characters and chorus invoke in the agon indicate that Euripides' style also has low, urban features, while Aeschylus' is lofty and associated with deep-rooted rural ritual (e.g., 822–25, 836–39). Like the Initiates' hailing of the "seats" (ἕδραις [i.e., precincts]) of Iacchos, of Korê, and of Iacchos himself as "much honored" (*polytimêtos*, 324, 337, 398), before the agon Dionysus addresses Aeschylus as *polytimêtos*. This is in direct contrast to Euripides, whom he greets as "O mischievous one" (ὦ πονηρ'), an epithet that suits the younger poet's roguish (*panourgos*) type (851–52; cf. 921).[41] Good citizen of Eleusis that he is, Aeschylus then prays to Demeter, "nourisher of [his] mind" (ἡ θρέψασα τὴν ἐμὴν φρένα), that he might be worthy of her Mysteries (886–87).[42] Euripides offers instead a suitably newfangled prayer, one to the air on which he feeds, the "hinge of the tongue," and his "sharp-sniffing nostrils" (αἰθήρ, ἐμὸν βόσκημα, καὶ γλώττης στρόφιγξ, καὶ ξύνεσι καὶ μυκτῆρες ὀσφραντήριοι, 892–93).

Their very distinct prayers are fully in keeping with differences that the chorus and Euripides himself have already indicated between the light, urbane, feminine polisher of words (ἐπῶν βασανίστρια λίσπη, 826) and his weighty, "wild" elder (ἀγριοποιὸν αὐθαδόστομον, 837[43]). Later the chorus claims that Aeschylus composes "the greatest and most noble verses" (πλεῖστα / καὶ κάλλιστα μέλη) and deems him "the

41 This is noticed by both Lossau 1987: 236–37 and Lada-Richards 1999: 248–49.

42 Scholars have also puzzled over the suggestion in the scholia that Aeschylus revealed some of the secrets of the Mysteries (e.g., Bowie 1993).

43 On Aeschylus' "sound-portrait" see Scharffenberger 2007.

Bacchic lord" (τὸν βακχεῖον ἄνακτα) (1254–59), associating his fulsome modes and raging demeanor with the god's inspired frenzy.[44] Euripides treats this wild style with disdain, declaring it weighty and overblown (οἰδοῦσαν ὑπὸ κομπασμάτων καὶ ῥημάτων ἐπαχθῶν, 940). This follows on a claim that Aeschylus' tactics aim at deceiving spectators who were already "fools raised in the school of Phrynichus" (μώρους λαβὼν παρὰ Φρυνίχῳ τραφέντας, 910), thus impugning older poetic styles as fostering simple-minded citizens.[45] Elsewhere in Aristophanes Phrynichus, Aeschylus' poetic predecessor, represents an earlier mode associated with rural spaces (e.g., *Vesp.* 220, *Av.* 748–50). Later in *Frogs* Aeschylus will himself invoke Phrynichus in order to oppose his own "pastoral" style – using a famous metaphor from archaic lyric of the poet as a bee feeding in the meadow (λειμῶνα, 1300)[46] – to Euripides' trashy urban Muse (1298–1303).

Euripides, in this scheme, is quintessentially citified: he chatters with Socrates in the Agora and encourages young men to do the same (1068–71, 1491–92). He is also a "twister" (892, 957) who slims down the tragic Muse (937–47), favoring a slender, "peripatetic" style (cf. the diet, e.g., ἐπυλλίοις καὶ περιπάτοις, 942). We might recognize, however, that this mobile, slimming mode is the urbane sister to erotic pastoral and the marshy modulations of the Frogs. The agon features a set of embodied styles that reconfigure and supplement the distinctions established along the path to Hades. Now Aeschylus emerges as not only of the Mysteries' meadows but more belligerently "natural," while the Frogs' "many-diving melodies" that recall Euripides' choral odes are capped by his light, feminized, and overly modulated monodies, as these are lampooned by Aeschylus.

44 See Lada-Richards 1999: 242–47; Bakola 2010: 26. This is, of course, one of the many places where Aristophanes echoes the *Bacchae*.

45 See, e.g., Rosen 2006; Worman 2008: 110–15.

46 Note that Aeschylus emphasizes his pastoral innovation: this is *not* the same meadow as that of his predecessor Phrynichus. Cf. Bacch. 10.10; P. *P.* 10.53–54; Ar. *Birds* 748–50; Pl. *Ion* 534b.

This is a complicated scheme. Euripides would appear to be both a "fertile," outlandish poet and yet one very much versed in the close-shaving modes of the city. Elements from Mysteries rituals and pastoral styles seem to converge especially upon the figure of Aeschylus, and this should dovetail nicely with the somewhat positive aesthetic judgment that the older poet eventually achieves. Yet most scholars of Aristophanes' literary critical and paratragic gestures notice that he is more interested in (and himself more stylistically similar to) Euripides.[47] Both younger poets are much more engaged with the discourse of literary criticism than is Aeschylus; but this discourse is itself tied in poetic convention to pastoral settings. While the association between the Frogs' imagery and rhythms and those of Euripides conflicts with the fact that later in the play Euripides is clearly figured as an urban type, it is also the case that *Frogs* frames rural spaces and their denizens as "rural" – appropriated by the urban discourse as metaphors or metonymies for critical interventions within the civic sphere. And yet the symbolic terrain that the play charts has its suggestive referents in actual places along the urban-rural divide. A closer look at these may aid further in understanding how to sort out the critical distinctions.

THEATRICAL SPACE AND RITUAL PLOTTING

While all of Aristophanes' extant plays reveal his bold and flexible use of mimetic space, *Frogs* achieves a shifting from place to place that is quite singular. In the first half of the play the mimetic scheme alters continuously, as Dionysus makes his way through marsh and meadow to Hades, the scene of poetic judgment.[48] The stage space is thus repeatedly overlain with

47 As Cratinus' coinage εὐριπιδαριστοφανίζειν (*inc. fab.* fr. 342) suggests; see Biles 2002, Ruffell 2002. Cratinus' tragic associate is, of course, Aeschylus; see Bakola 2010: 24–29 and further in section 4.

48 Russo (1962) 1994: 209–13 notes that this journey is unique in extant comedy and that it involves a continuous relabeling of the stage space.

adjacent fictional landscapes; and this changing topography brings together ritual practice, spatial "progress," and fitful glimpses of a literary landscape that highlights essential stylistic differences between the tragic poets.

Although readers of *Frogs* often disdain arguments that the play may also map actual spaces leading out of the city, the vibrant landscape details clearly point in this direction as well.[49] That said, the course that Aristophanes traces bears relatively few indications of Eleusinian topographies, the path that many scholars have favored. This journey would follow the road from Athens to Eleusis, and thus a route out of Athens along the Sacred Way (*Odos Hiera*) to the northwest and across the Cephisus River. The Frogs' altercations with Dionysus may be reminiscent of the "bridge insults" hurled at Initiates as they passed over the Cephisus, although the dramatic dynamic does not match very closely what we know of the ritual.[50] The choruses of Initiates (Mystai) also support the importance of Eleusis; but they are surrounded by elements that urge attention to the confluence of ritual and literary engagement in local spaces.[51] More significantly, the marsh scene bears no apparent relation to the Eleusinian Mysteries, while it does evoke a local celebration.

Thucydides identifies the area south of the Acropolis as part of the oldest ancient settlement and locates there the sanctuary of Dionysus Limnaios and the Anthesteria (*Hist.* 2.15–16). In the classical period this area along the Ilissos was a suburban

49 For arguments in favor of this scheme see Gerhard 1858; Tucker 1904; Hooker 1960; Guarducci 1982; Slater 1986. Tierney 1934, Segal 1961, and Dover 1993a offer support for the prominence of local rituals in the play but do not think a particular topography is indicated, although many of Dover's remarks emphasize Eleusis.

50 See Richardson 1974: 214; O'Higgins 2003: 57. But cf. Tierney 1934: 200–01, who points out that the Initiates themselves engage in insults, rather than serving as targets; Segal 1961: 235, who associates the *gephurismos* with the country (Lenaean) Dionysia; also Burkert (1977) 1985: 287, who thinks the "bridge insults" occurred closer to Eleusis. Cf. Edmonds 2004: 126n. 36.

51 See also Moorton 1989; Edmonds 2004.

district that was home to many temples, dedicatory altars, and shrines, including those to the Muses, Aphrodite, Pan, and the Nymphs.[52] It also housed springtime rituals and rites of passage, such as those that took place at the precinct of Demeter and Persephone at Agrai and possibly that of "Aphrodite in the Gardens," although determining the site of the latter has proved difficult.[53] The *Frogs'* trip through the marshes and to the flowery meadows of "Hades"[54] repeatedly evokes the fertility rituals of the Lesser Mysteries and the Anthesteria, which included rites at the sanctuary of Dionysus Limnaios, as well as aspects of the Lenaean festival itself, in which the play was performed.[55]

Scholars have interpreted the topography charted in mimetic space onstage in various ways, while some disdain the idea that Aristophanes depicts anything more particular than "symbolic space" (e.g., flowering meadows = paradise). But were we to attempt to follow an actual route, it would look something like this. Dionysus visits Heracles at his "house," either at Kynosarges, just outside of the south walls and close to the Diomeian Gate, or farther to the north, at the temple of Heracles

52 On Pan in Athens, see Borgeaud (1979) 1988: 140–41, 151–56; on Nymphs, Dionysus Limnaios, the Ilissos, and Agrai, see Larson 2001: 127–29.

53 For some of the relevant details see Wycherley 1963, 1978; and further below.

54 While Rogers 1919: xvi went so far as to characterize the *katabasis* scenes as "loosely tacked on" (cf. Russo (1962) 1994: 206–08), Segal 1961: 236–37 argues for connections between cultic practice and the dramatic festivals, most notably the Lenaea; in this he follows Tierney 1934. The theme of descent may itself augment the interconnections between pastoral spaces and poetry, since both Dionysus and Orpheus – the pastoral poet par excellence – make journeys to Hades.

55 These occurred in late February and involved the drinking of "pitchers" (*choes*), some seasonal appreciating (i.e., of the new vintage and the transition to the festival of flowers [the Anthesteria proper]), and offerings of pots (*chytroi*) of seeds to the dead. See Segal 1961: 219–20; Moorton 1989: 314–16; Hamilton 1992; Bowie 1993. On the connections to the Lenaea, see especially Tierney 1934: 206–12.

Alexikakos on the eastern slope of the Hill of the Nymphs.[56] He then proceeds to his own sanctuary "in the marshes," which some scholars thought referred to an area between the Pnyx and the Acropolis but that most locate close to the Ilissos.[57] Dionysus successfully crosses these marshes (*limnai*), accompanied by Frogs who invoke the drunken celebration of the Anthesteria. After this crossing, he arrives at flowering meadows for the "play" associated with Lesser Mysteries rituals, amid blooms suggestive of the Anthesteria.[58] Finally, he knocks on the door of "Hades," perhaps at Agrai, an area on the south side of the Ilissos that housed the temple of Demeter and Korê/Persephone, where the Lesser Mysteries were celebrated and where some scholars locate the Lenaean theater.[59] Others doubt

[56] For Kynosarges, see DL 6.1.13. Wycherley 1978: 229–30 discusses possible locations and the site's use as a gymnasium; see also Judeich 1931: 419; Hooker 1960. A scholion to *Ran*. 501 locates the temple of Heracles Alexikakos in Melite, a district that some think included the eastern slope of the Hill of the Nymphs (see Tucker and Harrison 1904: 418; Wycherley 1959; Lalonde 2006: 86–87).

[57] This is in keeping with Thucydides' placement of the sanctuary to the south of the Acropolis (πρὸς νότον, *Hist*. 2.15–16), and no other area around Athens was known as marshy. He also identifies this area as the setting for the "Older Dionysia" in the month of Anthesterion; Ps.-Demosthenes (*Ag. Neaera* 76) confirms the temple's antiquity and its connection to the Anthesteria. See Bates 1899; Tucker and Harrison 1904; cf. Tierney 1934: 204. Also Hooker 1960; Dover *ad* 216f.; Slater 1986. Wycherley 1963, 1978: 169–74 charts Socrates' path in the *Phaedrus* as slightly upriver from this area.

[58] As opposed to the Eleusinian rituals, which took place in the fall, when meadows would be dry.

[59] Travlos 1971 places the temple of Demeter and Kore in Agrai, across the Ilissos above Kynosarges. Slater 1986 thinks the *Frogs* journey ends at the Lenaean Theater in Agrai and has more to do with comic rituals. This has proved impossible to determine; while it is not even clear that there was a separate theater, many readers of the evidence think that there must have been and that it would have been either in the Agora (e.g., Tierney 1934: 204) or outside the city walls. It may also be relevant that half of the responsibility for running the Lenaea fell to the same officials that oversaw the Mysteries, which indicates at the least that the socioeconomic arrangement assumes an institutional coherence between the comic *agôn* and the rites of Demeter (Dover 1993b: 183). In addition, Tierney points to evidence

the existence of a separate theater, however, and most assume that *Frogs* was performed in the Theater of Dionysus.[60]

If Aristophanes did in fact present *Frogs* on the south slope of the Acropolis, and this terrain along the Ilissos is at least one of those suggested by the play's action, then it is alone in having a curious meta-dramatic significance. It would have been visible in the distance to audience members seated in the higher seats of the Theater of Dionysus, which looked out over the city walls toward the Ilissos and Agrai – again, toward the Athenian district where the Lesser Mysteries and the Anthesteria were celebrated. Since *Frogs* was produced for the Lenaea of 405, and thus in the winter festival that fell between the Lesser Mysteries and the Anthesteria, it seems likely that the journey onstage is meant to intersect with the ritual settings visible beyond the walls.[61] The Dionysus of Aristophanes' play would thus be guided onstage along the local route that these celebrations would take. For the original audience, then, the play effected on one level at least a witty conflation of mimetic, ritual, and actual space, implicating the various Dionysiac rituals across one variegated landscape.

The connections between the Frogs' imagery and that of Euripides' choral odes introduce an additional ritual constellation, one also connected to fertility, but via girls' rites of passage. While we should not make too much of the Euripidean echoes in Aristophanes' vocabulary and token deities, the Frogs' strophes do deploy elements of Euripides' choral odes that depict "pastoral" song and dance involving young women in transition, young women dramatized by Euripides whose plights resemble Persephone/Korê, whose story was central to the Mysteries. The Mysteries thus intersect with bucolic imagery in lyric poetry through their core legend, that of a girl picking flowers in a meadow, getting snatched by a god,

that the Lenaea was ritualistically bound up with the Dionysus of the Mysteries (Tierney 1934; also Segal 1961: 219–20).

60 Again, see Slater 1986; vs. Pickard-Cambridge 1927; Stanford 1957, 1958: x–xi; Segal 1961: 239–40n. 75.

61 See especially Moorton 1989: 316.

descending to the Underworld, and eventually establishing the cycle of seasons.[62] Further supporting the idea that a local ritual constellation is being engaged here, Philippe Borgeaud has argued that the Athenian Pan – which, again, Euripides' odes celebrate – has important ties to the rites of Demeter.[63]

The presence of elements that recall archaic lyric and perhaps especially eroticized pastoral spaces not only suggests an association of Euripides' modes with these spaces but also crafts a connection between that style and a particular aspect of the Eleusinian Mysteries: Persephone's "plucking the flower."[64] As Calame notes, this kind of imagery is central to archaic lyric and indicates passage through the "meadow" of young girls' burgeoning sexuality into the civic sphere.[65] An additional presence in this variegated topographical scheme may further connect ritual landscapes and what Calame calls "the metaphorical spaces of love." Most tellingly, the sanctuary of Aphrodite "in the Gardens," which some scholars locate on the banks of the Ilissos, had ritual ties to that of Pandrosus (on the north side of the Acropolis next to the Erechtheion) in the form of practices regulating young girls' fertility that included a nocturnal descent into the garden precinct.[66]

Within the larger context of pastoral (or "pastoral") scenes of inspiration and judgment, this area along the Ilissos is unique in its association with both fertility rituals and elements related to poetry and aesthetic judgment. Not only does

62 Cf. Calame 1999: 153–74, 2007. The monodies of Helen (*Helen*) and Creusa (*Ion*) depict their rapes as from among the flowers, next to waterfalls. The imagery of the Frogs' song contains some echoes of the stasimon in *Ion* in which the chorus depicts the rape of Creusa.

63 Borgeaud (1979) 1988: 141–50.

64 Cf. also Eur. *Med.* 824–44, 1085–89, *Bacch.* 395–420, *IA* 543–67.

65 E.g., Archil. fr. S478a-b [Cologne Epode]; Anacr. frs. 346, 417 P; Sapph. frs. 96, 105b P. See Calame 1999: 165–67. On the gendering of ritual space, see Cole 2004.

66 Paus. 1.27.2–3. See Calame 2007 on "gardens of love;" also Calame 1999: 170–74. There are many problems with both the understanding of the rituals and the situation of the sanctuary; see Wycherley 1963: 94, 1978: 172; Travlos 1971: 228; Motte 1973: 121–37; Brulé 1987: 84–90.

it house sites dedicated to deities connected to poetic production in rustic settings, including an altar to the Muses in their guise as "daughters of Ilissos" (Paus. 1.19.6); it is also one of the famous sites that accrue significance in the literary critical tradition.[67] To put it in schematic relation to the spaces and their poetic inhabitants: the comic Dionysus fords the marshy expanse to the beat of a Frog chorus who claim a special role in the fostering of a bucolic mode; he then witnesses the chorus of Initiates singing among and about flowering meadows that are traditional metonyms for erotic pastoral modes in lyric poetry.[68] Finally, he enters Hades somewhere out there in the countryside, the nether realm associated with the nature poet Orpheus, where the tragic poets do battle at least in part by situating themselves along the urban-rural divide.[69] Further, *Frogs* employs semi-bucolic ("suburban") settings and more fully rural landscapes, both of which host Athenian fertility rituals and intersect with the route to the testing of tragedy. That is, these spaces and their details arbitrate between city and country modes, between Euripides and Aeschylus, and ultimately between tragedy and comedy. We might, then, recognize in *Frogs* an innovative appropriation of bucolic settings as a means of forging a new and distinctly hybrid critical

67 Others are Helicon and its Valley of the Muses, Olympus' Pierian vale, Olympia, the Acropolis and Hymettos caves of Pan and nymphs; see Worman forthcoming.

68 Cf. esp. Sappho fr. 2 L-P; Anacreon fr. 417 P; Ibycus fr. 286 P-D; *HDem*. In *Frogs* this conventional space also effectively conjoins aspects associated with Euripides (girls' songs) with those associated with Aeschylus (meadowland settings).

69 See Sperduti 1950: 213–17; Segal 1961: 221, 1989: 8–10, who points to the Hellenistic roles of Orpheus and Dionysus as fertility deities; as well as Bowie 1993: 230–34, who emphasizes the similarity in Eleusinian and Orphic settings. In the case of Orpheus, this aestheticizes earlier Orphic writings that depict the poet's song as calming and ordering nature, what Segal 1989: 10 calls, "a resonant harmony." Although *Frogs* does not explicitly refer to Orpheus, narratives of *katabasis* inevitably recall him, as Edmonds 2004: 10 has argued, following Tierney 1934: 218 and n. 74 and Bowie 1993: 230–32. See also Whitman 1964: 228–31 and Padilla 1992 on the relevance of Heracles to such themes.

rubric – call it "fertile" judgment. From this perspective, a central means for assessing poetic worth would involve judgments about its capacity for civic renewal through rural rites, something perhaps more closely connected to the older fertility rituals of which the Lenaean festival is an offshoot.

Thus the path through the marshes and fields outside of the city walls would also emerge as the proper "path" or mode of judgment, given that it ultimately gets the journeyer to the space and perspective from which he can hope to assess clearly what the city needs. One further factor may account for this emphasis on the rural or pastoral setting: the tendency in Attic comedy to send protagonists out into the countryside to fix what has gone wrong with the city. Aristophanes seems to have made the most frequent use of this pattern, which is especially evident in *Acharnians* and *Birds* but serves as a background for understanding *Knights*, *Peace*, and *Wealth* as well. When Aristophanes launches his assessments of Athens' ills, he usually opposes old, rural, bean-chewing ways (cf. *Eq*. 41, 805–08) to newfangled, louche, urban ones. Think, for instance, of the pastoral idyll in the Academy that the Stronger Logos offers dreamily in *Clouds* (1002–08), versus the urbane symposium for which Bdelycleon tries to gussy up his rustic (*agroikos*) father in *Wasps* (1131ff.; cf. 1320).[70] Such contrasts suggest that, despite its denizens' much-lampooned crudeness, the country may offer an older, more morally grounded context from which to judge the city's foibles, both aesthetic and ethical.[71]

Dionysus would, from this perspective, represent a version of the typically rustic comic protagonist whose unsophisticated literary tastes match his recourse to rural retreats for information and resolution. Things are not quite this simple, of course: Dionysus initially has an aesthetic crush on the urbane Euripides; and he is not very obviously cut in the "old

[70] Wycherley 1978: 220 notes Aristophanes' portrait and emphasizes the Academy's park-like aspects.

[71] Cf. also the staunch critique of the "rural" Dicaeopolis in *Acharnians*; see Rosen 2006 on the mockery of *agroikia*.

country" mold. But to the extent that Dionysus' character is more hybrid than the typical buffoonish hero, he matches in semiotic complexity both the critical idiom and its representative, Euripides. Since the resolution of *Frogs* must be achieved in literary critical (as opposed to, say, strictly political) terms and since literary convention had already appropriated semi-rural settings as "pastoral" – that is, as sites for the culling of poetic imagery for what are essentially urban (i.e, civic cultural) uses – it makes some sense that this poetry-loving protagonist would take this trek outside the city walls in search of a solution to his problem.

WHAT RIVERS CAN MEAN

While comic imagery would seem to associate critical practices with Euripides, the Agora, and city activities, in *Frogs* Aristophanes places them and the poet in the countryside. And, again, one of the routes indicated seems to pass along the Ilissos, the riverside setting that Plato uses to frame the action of the *Phaedrus*, a dialogue that appropriates the imagery of poetic inspiration and "nympholepsy" for use in judging speeches about love.[72] Like Aristophanes, Plato wryly depicts this quasi-pastoral setting as enhancing artistic sensibility and thus as a spur for and even necessary to the process of critical judgment. While both comic poet and philosopher treat their pastoral topographies with an engaged and delighted irony, they also effectively establish such spaces as discursively essential. Later theorists do not miss this emphasis: for Cicero and Dionysus of Halicarnassus in particular, the details of the Ilissos setting are symbolic of and provide crucial distinctions for the terms of critical judgment.[73]

Further, the fact that the *Phaedrus* is staged in this same general area suggests that the highlighting of this quasi-pastoral

72 For the setting see Wycherley 1963; Lind 1987. On the placing of nympholepsy see Connor 1988.

73 Cic. *De orat.* 1.28–29, 2.12–24; *De leg.* 1.1–4; *De fin.* 5.1–9; DH *Dem.* 5.5–17.

space, with its shrines to nymphs and the Muses, may serve as a fitting frame for the developing contentions among literary critics, rhetoricians, and philosophers about which practices have or ought to have civic benefits and which involve a retreat from the city and its corrupting concerns. Poetry may in comic representation still be called upon for its civic wisdom, but the practices by which its own value is assessed are faulty and in need of rustic correction. The reinvigoration of poetry may demand a return to country scenes of inspiration and authority, but in Aristophanes and Plato alike these settings have very ambiguous effects on critical judgment. Euripides and Socrates both bridge this divide, since they practice the deflationary, close-shaving measures of the Agora and yet bring into play the feminized spaces of ritual and erotic enactment beyond the city walls.

Landscape features such as rivers underscore these distinctions, charting stylistic continua that run from fluid and fertile to rough and roiling. A river's smooth flow can suggest a mode quite dissimilar to other bodies of water; and the Ilissos itself indicates a range of characteristics within the stylistic scheme, sometimes linking poetic modes with a watery fertility, at other times indicating excess. In an arid climate, water really is best, in that it alone gives life to dry land; but in the realm of poetry and literary criticism this fluid bounty is recognized as potentially too much of a good thing. Poets and critics often deploy water imagery to point up a fluidity of phrasing and word order, which may be like a clear or smooth and flowing stream (usually welcome) or a full and rushing river or flood (impressive but overwhelming).

In the last quarter of the fifth century the distinction between the great flood and a more restrained mode centers, at least in part, on an aesthetic and ethical debate between Aristophanes and his senior comic poet Cratinus; and it combines the notion of the big flood with drinking.[74] Aristophanes

[74] For the reception of this set of distinctions, see, e.g., Crowther 1979; Dunn 1989.

depicts his elder poet as marked by an unleashed, overly flowing style:

> εἶτα Κρατίνου μεμνημένος, ὃς πολλῷ ῥεύσας ποτ' ἐπαίνῳ διὰ τῶν ἀφελῶν πεδίων ἔρρει, καὶ τῆς στάσεως παρασύρων ἐφόρει τὰς δρῦς καὶ τὰς πλατάνους καὶ τοὺς ἐχθροὺς προθελύμνους.
>
> Then [he says that he] remembers Cratinus, who flowing with abundant praise, flooded the open plains and, sweeping them from standing, carried off oaks and elms and enemies, roots and all. (Ar. *Knights* 526–28)

Aristophanes joins this image to that of a gabbling, bibulous old age, so that "flooding" and "drinking" (i.e., wine in excess) are stylistically aligned. Scholars have taken a fragment from Cratinus' *Putinē* (*Wine Flask*) to be a rambunctious, self-mocking response to this unflattering portrait. Here the poet depicts a character overwhelmed by this superabundance of words:

> Αναξ Ἄπολλον, τῶν ἐπῶν τῶν ῥευμάτων.
> καναχοῦσι πηγαί, δωδεκάκρουνον τὸ στόμα,
> Ἰλισσὸς ἐν τῇ φάρυγι· τί ἂν εἴποιμί σοι;
> εἰ μὴ γὰρ ἐπιβύσει τις αὐτοῦ τὸ στόμα,
> ἅπαντα ταῦτα κατακλύσει ποιήμασιν.
>
> Lord Apollo, fountains of flowing words
> splash out, his mouth has twelve springs,
> an Ilissos in his throat – what can I say?
> If someone doesn't stopper his mouth,
> he will flood the whole place here with verses.
> (Crat. fr. 198 K-A)

Cratinus famously claimed that drinking water alone would foreclose any creativity (fr. 203 K-A), a contrast that seems to have enjoyed a piquant afterlife among the orators, especially in criticisms of Demosthenes' careful, polished style.[75]

[75] See Worman 2008: 108, 248–49.

This volubility presumably contrasts with Aristophanes' restrained orifice (στόμαθ' ἡνιοχήσας) at *Wasps* 1022, the drama that he produced in the following year and that may respond at least in part to Cratinus' *Putinê*.[76] With his drunken creativity and flood of words, Cratinus resembles Archilochus, whose stylistic heir he claimed to be and whose reception echoes this association. But *Frogs* offers us another roiling poet, in the form of Cratinus' own favorite tragic target Aeschylus, whom the chorus encourages in the *agôn* "boldly to release [his] fountain" (θαρρῶν τὸν κρουνὸν ἀφίει, 1005).[77]

The marshy riverside setting in *Frogs* seems to be in dialogue with this tradition, offering a new turn on what the "watery" style might be. If the bibulous, full-flowing associate of Aeschylus can boast that he has an Ilissos in his throat, perhaps the slighter mode of the Frogs, of Euripides – and indeed of Aristophanes – can lay claim to a different aspect of this river's fluid aspects. Although late in the drama Aeschylus' addition of "river" (ποταμέ, 1383) to the line he enters upon the scales of poetry gives it weight, as Dionysus explains (ὅτι εἰσθέθηκε ποταμόν, 1386), this may not be the kind of fertile source the discerning critic is after.[78] The early scenes of *Frogs*, like the opening of Plato's *Phaedrus*, suggest lighter associations for the Ilissos and its aesthetic qualities: again, that of

76 Biles 2002: 189–201 focuses especially on the figure of Philocleon, in whose drunken behavior at symposium he sees a lampoon of Cratinus. See also Rosen 2000; Lada-Richards 2002: 85; Ruffell 2002; Bakola 2010: 20–33.

77 This anticipates Callimachaen aesthetics, of course (e.g., *HAp.* 110–12). Cf. also Archil. fr. 120 W regarding the creative powers of wine, as well as, again, Cratinus' dismissal of the creative potential of the water drinker (fr. 203 K-A). The testimony 17 K-A seems to suggest that Cratinus consciously fashioned his poetic persona and style after Archilochus. Ps.-Longinus (33.5) deprecatingly compares Archilochus' style to an unleashed flood, an image that the careful and restrained Horace in his satires uses to denigrate his predecessor Lucilius' style (*Serm.* 1.4.21, 1.10.50).

78 Lada-Richards 1999: 242–47, 2002: 85 aligns *gonimos* with Cratinus' and Aeschylus' "flood" of words; but the Aristophanic aesthetic revises such equations by associating Euripides and a lighter style with fertile modes.

the sparkling, fine-flowing stream versus the roiling river. The waters of *Frogs*, insofar as they may toy with the actual setting of the Ilissos, foreground the marshy qualities of a river overflowing its banks in springtime, while the gentle, cooling waters of the *Phaedrus* setting clearly describe the Ilissos in the heat of summer, when its waters would be lower and easily fordable (cf., e.g., *Phdr.* 238c5-d3). But neither the play nor the dialogue emphasize the thunderous qualities of a rushing river; rather, in both, watery elements indicate fluid, agile styles. Within this scheme Euripides' style falls on the side of light-flowing brooks and smooth streams, while Aeschylus' is embodied by the full and weighty river that he so proudly offers up in his winning line.

Aristophanes' drama of the path to Hades for the judgment of poets sets up a critical frame that is site-specific and ritually infused. As Dionysus takes his trip in search of the dexterous and fertile Euripides, both his poetic modes and those of Aeschylus accrete the tactile qualities of topographies inhabited by celebrants of the ritual spaces through which the god passes. In this way, the critique builds as the journey progresses; and while Dionysus remains irresolute, easily distracted, and taken in by cheap theatrics, the literary critical process itself emerges as one that clarifies distinctions between poetic styles by mapping ritual processions in significant spaces along the path out of the city. Perhaps it is only by means of the "distance" (conceptually and within mimetic space) created by the trip through the marshes and meadows that the proper literary judge can emerge: the comic *didaskalos*, with his quasi-rural roots, connections to suburban fertility rituals, and hybrid critical discourse. If poetic contests reveal themselves as flawed, as often based on silly criteria and assessed by foolish, ignorant judges, the play gestures toward the possibility of more discerning critical practices (e.g., well-"grounded" literary judgment) that could offer civic enrichment and renewal.

Plato, great admirer of Aristophanes that he was, highlights the attraction of such rural correctives, when he envisions in the *Republic* how young citizens' constant exposure to the overabundance of art forms that characterize the lively city may be cured by a trip into the countryside. Socrates uses a "healthy meadow" topos to argue that the young should be "pastured" in a place where they will be exposed only to good impressions. Thus like well-tended cattle they will not run the risk of "grazing" on bad images coming from many different sources (ἵνα μὴ ἐν κακίας εἰκόσι τρεφόμενοι ἡμῖν οἱ φύλακες ὥσπερ ἐν κακῇ βοτάνῃ, πολλὰ ἑκάστης ἡμέρας κατὰ σμικρὸν ἀπὸ πολλῶν δρεπόμενοί τε καὶ νεμόμενοι). Instead, as with breezes that blow from a healthy setting, the young citizens will be nurtured from all directions if they are surrounded by examples of good actions, which may waft over them visually or aurally (ὥσπερ ἐν ὑγιεινῷ τόπῳ οἰκοῦντες οἱ νέοι ἀπὸ παντὸς ὠφελῶνται, ὁπόθεν ἂν αὐτοῖς ἀπὸ τῶν καλῶν ἔργων ἢ πρὸς ὄψιν ἢ πρὸς ἀκοήν τι προσβάλῃ, ὥσπερ αὔρα φέρουσα ἀπὸ χρηστῶν τόπων ὑγίειαν) (*Rep.* 401b1–d2). In keeping with the moral topography of *Frogs*, Socrates' analogies place the finer, nobler civic values and modeling for the young outside of the city, so that the rustic imagery configures a purer, simpler space and time. This fantasy space, which *Frogs* frankly recognizes as such, Plato takes up as a future ideal.

Aristophanes' literary critical journey to the space of judgment enacts a fusion of dramatic and religious rituals to renovate an old poetic convention that situates inspiration and authority in rural settings. It thereby fosters a long tradition in philosophy and rhetorical theory of the political, ethical, and aesthetic refreshment that this trip out of the city affords. Further, it grounds in concrete and influential ways an emerging set of distinctions between grand, imposing styles and light, subtle ones. Hellenistic and Roman writers elaborate on these stylistic differences frequently by supplementing classical landscape details with some of their own, so that rural imagery continues to signal delicacy at one end (e.g., flowering gardens, pure springs) and grandeur at the other (e.g., rough mountains, roiling rivers). Aristophanes' emphasis on these pastoral spaces also encourages among the later critics the insight that natural settings may offer the most vibrant indicators of stylistic differences. What *Frogs* establishes, then, is not only that proper critical judgment necessitates a retreat from urban spaces, but also that all styles may have their rural roots. Thus even Euripidean dash and polish share features with the sparkle and flow of the river's stream, while the styles of older poets like Aeschylus and Phrynicus – that old meadowland bee – hail from farther afield.

Bibliography

Asper, M. 1997. *Onomata allotria: Zur Genese, Struktur und Funktion poetologischer Metaphern bei Kallimachos*. *Hermes Einzelschriften* 75. Stuttgart.

Bakola, E. 2010. *Cratinus and the Art of Comedy*. Oxford.

Bates, W. N. 1899. "The Lenaea, the Anthesteria, and the Temple en Limnais." *TAPA* 30: 89–98.

Becker, O. 1937. *Das Bild des Weges und verwandte Vorstellungen im frühgriechischen Denken*. *Hermes Einzelschriften* 4. Berlin.

Biles, Z. P. 2002. "Intertextual Biography in the Rivalry Between Cratinus and Aristophanes." *AJP* 123.2: 169–204.

Borgeaud, P. (1979) 1988. *The Cult of Pan in Ancient Greece*, trans. K. Atlass and J. Redfield. Chicago.

Bowie, A. M. 1993. *Aristophanes: Myth, Ritual, and Comedy*. Cambridge.

Brulé, P. 1987. *Le fille d'Athènes. La religion des filles à Athènes à l'époque classique*. Besançon and Paris.

Burkert, W. (1977) 1985. *Greek Religion*, trans. J. Raffan. Cambridge, MA.

Buxton, R. 1994. *Imaginary Greece: The Contexts of Mythology*. Cambridge.

Calame, C. 1999. *The Poetics of Eros in Ancient Greece*, trans. J. Lloyd. Princeton.

2007. "Longus and the Pastoral Garden of Eros." In *Sacred Gardens and Landscapes: Ritual and Agency*, ed. M. Conan. Washington, DC. 43–54.

Campbell, D. A. 1984. "The Frogs in the *Frogs*." *JHS* 104: 163–65.

Cole, S. G. 2004. *Landscapes, Gender, and Ritual Space: The Ancient Greek Experience*. Berkeley.

Connor, W. R. 1988. "Seized by Nymphs: Nympholepsy and Symbolic Expression in Greece." *CA* 7.2: 155–89.

Cosgrove, D. and S. Daniels. (1988) 2002. *The Iconography of Landscape: Essays on the Symbolic Representation, Design and Use of Past Environments*. Cambridge.

Crowther, N. B. 1979. "Water and Wine as Symbols of Inspiration." *Mnemosyne* 32.1–2: 1–11.

Csapo, E. 2004. "The Politics of the New Music." In *Music and the Muses: The Culture of Mousikê in the Classical Athenian City*, ed. P. Murray and P. Wilson. Oxford. 207–48.

Curtius, E. 1953. *European Literature and the Latin Middle Ages*. New York.

Defradas, J. 1969. "Le chant des Grenouilles: Aristophane critique musical." *REA* 71: 23–37.

Dover. K. J. 1993a. *Aristophanes: Frogs*. Oxford.

Dover, K. J. 1993b. "The Chorus of Initiates in *Frogs*." *Aristophane. Entretiens sur L'Antiquité Classique*, Foundation Hardt, vol. 38. Genève. 173–93.

Dunn, F. M. 1989. "Horace's Sacred Spring (Ode, I.1)." *Latomus* 48: 97–109.

Edmonds, R. G. 2004. *Myths of the Underworld Journey: Plato, Aristophanes, and the 'Orphic' Gold Tablets*. Cambridge.

Elam, K. 1980. *The Semiotics of Theatre and Drama*. London.

Elliger, W. 1975. *Die Darstellung der Landschaft in der griechischen Dichtung*. Berlin.

Fantuzzi, M. 2006. "Theocritus' Constructive Interpreters, and the Creation of a Bucolic Reader." In *Brill's Companion to Greek and Latin Pastoral*, ed. M. Fantuzzi and T. Papanghelis. Leiden.235–62.

Foley, H. P. 1994. *Homeric Hymn to Demeter: Translation, Commentary, and Interpretive Essays*. Princeton.

Franklin, J. C. 2013. "Song Benders of Circular Choruses: Dithyramb and the Demise of Music." In *Dithyramb in Context*, ed. P. Wilson and B. Kowalzig. Oxford. 213–36.

Gerhard, E. 1858. "Ueber den Iakchozug bei Aristophanes." *Philologus* 30: 210–12.

Guarducci, M. 1982. "Le *Rane* di Aristofane e la Topographia Ateniese." *Studi in Onore di Aristide Colonna*. Perugia. 167–72.

Halperin, D. 1983. *Before Pastoral: Theocritus and the Ancient Tradition of Bucolic Poetry*. New Haven, CT.

Hamilton, R. 1992. *Choes and Anthesteria: Athenian Iconography and Ritual*. Ann Arbor.

Harriott, R. 1969. *Poetry and Criticism Before Plato*. London.

Hass, P. 1998. *Der locus amoenus in der antiken Literatur: Ze Theorie und Geschichte eines literarischen Motivs*. Bamberg.

Hooker, G. T. W. 1960. "The Topography of the *Frogs*." *JHS* 80: 112–17.

Issacharoff, M. 1989. *Discourse as Performance*. Stanford. (Orig. "Space and Reference in Drama." *Poetics Today* 2 [1981]: 211–24.)

Jones, P. 2005. *Reading Rivers in Roman Literature and Culture*. Lanham, MD.

Judeich, W. 1931. *Topographie von Athen*, 2nd ed. Munich.

Lada-Richards, I. 1999. *Initiating Dionysus: Ritual and Theatre in Aristophanes' Frogs*. Oxford.

2002. "Reinscribing the Muse: Greek Drama and the Discourse of Inspired Creativity." In *Cultivating the Muse: Struggles for Power and Inspiration in Classical Literature*, ed. E. Spentzou and D. Fowler. Oxford. 69–91.

Lalonde, G. 2006. *Horos Dios: An Athenian Shrine and Cult of Zeus*. Brill.

Larson, J. 2001. *Greek Nymphs: Myth, Cult, Lore*. Oxford.

Lefkowitz, M. 1963. "*Tô kai egô*: The First Person in Pindar." *HSCP* 67: 177–253.

Lind, H. 1987. "Socrates am Ilissos: IG I3 1 257 und die Eingansszene des Platonischen 'Phaidros.'" *Zeitschrift fur Papyrologie und Epigraphik* Bd. 69: 15–19.

Lossau, M. 1987. "Amphibolisches in Aristophanes' *Fröschen*." *RM* n.f. 130: 229–47.

Lowe, N. 2006. "Aristophanic Spacecraft." In *Playing Around Aristophanes: Essays in Honour of the Completion of the Edition of the Comedies of Aristophanes by Alan Sommerstein*, ed. L. Kozak and J. Rich. Oxford. 48–64.

McGlew, J. 2002. *Citizens on Stage: Comedy and Political Culture in the Athenian Democracy*. Ann Arbor.

Moorton, R. J. 1989. "Rites of Passage in Aristophanes' *Frogs*." *CJ* 84.4: 308–24.

Motte, A. 1973. *Prairies et jardins de la Grèce antique: De la religion à la philosophie*. Brussels.

O'Higgins, L. 2003. *Women and Humor in Classical Greece*. Cambridge.

O'Sullivan, N. 1992. *Alcidamas, Aristophanes, and Early Greek Stylistic Theory*. Hermes Einzelschriften 60. Stuttgart.

Padilla, M. 1992. "The Heraclean Dionysus: Theatrical and Social Renewal in Aristophanes' *Frogs*." *Arethusa* 25: 359–84.

Panofsky, E. 1930. *Herkules an Scheiderwege und andere antike Bildstoffe in der neueren Kunst*. Leipzig. *Studien der Bibliothek Warburg* 18.

Parker, L. P. E. 1997. *The Songs of Aristophanes*. Oxford.

Parry, A. 1957. "Landscape in Greek Poetry." *YCS* 15: 3–29.

Pickard-Cambridge, A. W. 1927. *Dithyramb Tragedy and Comedy*. Oxford.

Power, T. 2010. *The Culture of Kitharôdia. Hellenic Studies 15*. Washington, DC.

Pucci, P. 1977. *Hesiod and the Language of Poetry*. Baltimore.

Radermacher, L. 1954. *Aristophanes Frösche: Einleiting, Text und Kommentar*. Vienna.

Richardson, N. J. 1974. *The Homeric Hymn to Demeter*. Oxford.

Rogers, B. 1919. *The Comedies of Aristophanes*, vol. 5. London.

Rosen, R. M. 2000. "Cratinus' *Putine* and the Construction of the Comic Self." In *The Rivals of Aristophanes: Studies in Athenian Old Comedy*, ed. D. Harvey and J. Wilkins. London. 23–39.

2006. "Comic Aischrology and the Urbanization of *Agroikia*." In *City, Countryside, and the Spatial Organization of Value in Classical Antiquity*, ed. R. M. Rosen and I. Sluiter. Leiden. 219–38.

Rosenmeyer, R. 1969. *The Green Cabinet: Theocritus and the European Pastoral Lyric*. Berkeley.

Rosenmeyer, T. G. 1962. "Plato's Prayer to Pan." *Hermes* 90: 34–44.

Rosenmeyer, P. 2004. "Girls at Play in Early Greek Poetry." *AJP* 125.2: 163–78.

Ruffell, I. 2002. "A Total Write-Off. Aristophanes, Cratinus, and the Rhetoric of Comic Competition." *CQ* 52.1: 138–63.

Russo, C. F. (1962) 1994. *Aristophanes: An Author for the Stage*. New York.

Schama, S. 1995. *Landscape and Memory*. New York.

Scharffenberger, E. W. 2007. "*Deinon Eribremetas*: The Sound and Sense of Aeschylus in Aristophanes' *Frogs*." *CW* 100.3: 229–49.

Schönbeck, G. 1962. *Der locus amoenus von Homer bis Horaz*. Diss. Heidelberg.

Segal, C. P. 1961. "The Character of Dionysus and the Unity of the *Frogs*. *HSCP* 65: 207–42.

1989. *Orpheus: The Myth of the Poet*. Baltimore.

Silk, M. 2000. *Aristophanes and the Definition of Comedy*. Oxford.

Slater, N. W. 1986. "The Lenaean Theater." *ZPE* 66: 255–64.

2002. *Spectator Politics: Metatheatre and Performance in Aristophanes*. Philadelphia.

Snell, B. 1955. "Das Symbol des Weges." In *Die Entdenkung des Geistes. Studien sur Entstehung des europoaïschen Denkens bei den Griechen*, 3rd ed. Hamburg.

Sperduti, A. 1950. "The Divine Nature of Poetry in Antiquity." *TAPA* 81: 209–40.

Stanford, W. B. (1957) 1983. "Notes on Aristophanes *Frogs*." *Hermathena* 89: 65.

1958. *Aristophanes: Frogs*. Bristol.

Steiner, D. 1986. *The Crown of Song: Metaphor in Pindar*. London.

Tierney, M. 1934. "The Parodos of Aristophanes' 'Frogs.'" *Proceedings of the Royal Irish Academy* Section C 42: 199–218.

Travlos, J. 1971. *Pictorial Dictionary of Ancient Athens*. New York.

Tucker, P. G. 1904. "The Mysteries in the *Frogs* of Aristophanes." *CR*, 18.8: 416–18. With a postscript by J. E. Harrison.

Ubersfeld, A. 1977. *Lire le théâtre*. Paris.

van Gennep, A. (1909) 1981. *Les rites de passage: Étude systématique des rites*. Paris.

Whitman, C. 1964. *Aristophanes and the Comic Hero*. Cambridge, MA.

Williams, R. 1973. *The Country and the City*. Oxford.

Wilson, P. 1999. "The *Aulos* in Athens." In *Performance Culture and Athenian Democracy*, ed. S. Goldhill and R. Osborne. Cambridge. 58–95.

Woodbury, L. 1988. "The Poetry of Tongue and *Phrên* in the *Frogs*." In *Language and the Tragic Hero: Essays on Greek Tragedy in Honor of Gordon M. Kirkwood*, ed. P. Pucci. Atlanta, GA. 175–85.

Worman, N. 1999. "Odysseus *Panourgos*: The Liar's Style in Oratory and Tragedy." *Helios* 26.1: 35–68.

2008. *Abusive Mouths in Classical Athens*. Cambridge.

Forthcoming. *Dreams of Order: Landscape Aesthetics in Ancient Literary Criticism and Theory*. Cambridge.

Wycherley, R. E. 1959. "Two Athenian Shrines." *AJA* 63.1: 67–72.

1963. "The Scene of Plato's *Phaidros*." *Phoenix* 17.2: 88–98.

1978. *The Stones of Athens*. Princeton.

Zetzel, J. E. G. 1983. "Re-Creating the Canon: Augustan Poetry and the Alexandrian Past." *Critical Inquiry* 10.1: 83–105.

7

THE PERMEABLE SPACES OF THE ATHENIAN LAW-COURT

Alastair J. L. Blanshard

The Peculiarity of "Forensic Space"

Rhetoric is a genre that desires the general but can never forget the specific. On the one hand, it attempts to generate a set of rules, methodologies, and aesthetics that can be applied irrespective of time and place. It aims to offer a speech for each and every occasion. On the other hand, it is a genre that constantly reminds its practitioners to be responsive to the peculiarities of circumstance, personnel, and audience.[1] Every speech is different and needs to be tailored accordingly.

One of the by-products of this tension between the general and the specific is that it is easy to lose sight of the role of space in the practice of rhetoric. At best, rhetoric seems reactive to environment rather than a proactive shaper of it. Against this position, I want to illustrate some of the ways in which rhetoric impacted on the physical environment through the construction of purposefully created spaces that were designed to facilitate its functions. In particular, I propose to examine the role and function of the spaces that were carved out for the performance of forensic rhetoric.

1 See, for example, *Rhet. Ad Alex.* 1426b35–40, 1437a-b; Arist. *Rhet.* 1356a5–16, 1359b27–38.

The Athenian law-court occupies a privileged space in the politics of the city. The writer of the *Athenaion Politeia* spoke for many when he declared the rise of the jury-courts as one of the most significant steps in the democratization of the state.[2] Meeting between 175 and 225 days a year and involving large citizen juries of regularly hundreds of citizens and occasionally thousands, the law-court has no rival in the city for the volume and importance of its rhetorical output.[3] A study of the physical environment for the performance of Athenian forensic rhetoric provides a useful test case for examining the role of place in the creation of rhetorical space in the classical city as well as supplying a template for the study of space and other genres of rhetoric.

Rhetorical performances did not occur in a vacuum. Instead, they played out in a performance space that was deliberately created to operate outside the normal sphere of civic space and according to its own particular rules and protocols. We can see the effects of this peculiarity of space in Aeschines' speech, *Against Timarchus*, where Aeschines highlights that there is something special about the language required to be spoken in court.

At the start of his discussion of the illicit activities of the alleged male prostitute Timarchus, Aeschines purports to run into a problem. It turns out that it is impossible to describe the activities of Timarchus without using inappropriate language.[4] Aeschines' words threaten to prove to be as bad as Timarchus' deeds. Talking about Timarchus is a real problem and the speaker's reticence proves to be a recurrent theme throughout the speech as Aeschines battles to be explicit about the nature of Timarchus' crimes.[5] What makes Aeschines' inability to speak so extraordinary is that his case is based on the fact that he doesn't need to prove his allegations because Timarchus' crimes are apparently so notorious that they have become a

2 *Ath. Pol.* 9.1.

3 On the frequency of law-court meetings, see Hansen 1979.

4 Aeschin. 1.37–38.

5 For other examples, see Aeschin. 1.55, 70.

regular point of discussion. Time and again, Aeschines asks the jury to just remember the gossip that they have heard.[6] Aeschines portrays himself not so much as a prosecutor, but as a prompter. "It is no great matter to prosecute when affairs are commonly known (ὁμολογουμένων) – all one needs to do is engage the memory of the listeners," he declares.[7] Indeed, Timarchus' activities are so well known that Timarchus cannot even say words such as "wall" or "tower" in public without people sniggering and attributing to them obscene connotations.[8] In an extended section in praise of the power of rumor, Aeschines declares Rumor to be one of the greatest gods in the city and the ultimate purveyor of truth.[9] The gossip that the city has been sharing here takes on the mantle of divine fiat. It seems that everybody, everywhere, all-the-time is speaking about Timarchus' crimes in graphic detail.

Aeschines' professed squeamishness comes across as odd particularly when we consider that other genres had no problem discussing the activities that Timarchus was alleged to have performed. The greedy politician who submits to anal penetration is practically a stock figure of Attic comedy.[10] We see him invoked time and time again. Yet despite this prominence on the dramatic stage, Aeschines can't bring himself to speak about Timarchus' crimes. The problem seems to be the location of his speech. There is something about the propriety expected of the speaker in the law-court that silences him. In Aeschines' prudishness, we see the discursive effects of the translation of the discussion of Timarchus' activities to "forensic space." Here the rules of language do not operate quite as they do elsewhere. It might be fine in the Agora or the symposium or the comic theatre to talk openly and freely about Timarchus' crimes, but here in the *dikasterion* ("law-court"), it is another matter. Here the conventions are different.

6 See especially, Aeschin. 1.89–93, 129.

7 Aeschin. 1.44.

8 Aeschin. 1.80.

9 For the section in praise of rumour, see Aeschin. 1.127–30.

10 See, for example, Ar. *Knights* 876–80, *Clouds* 1093, *Eccl.* 110–13.

Of course, it is tempting to dismiss Aeschines' protestations as just rhetorical posturing, an all-too-convenient excuse by Aeschines to cover up his lack of evidence while attempting to make himself look like an upstanding moral citizen. Yet if we treat his statements as rhetorical gambits – and on some level we must – we shouldn't also downplay the ritualistic effect of such utterances. They operate to create a sense of space and occasion. There are elements of the meta-rhetorical in Aeschines' text. They signal to the audience that there is something peculiar about the place they inhabit.

As a number of other commentators have observed, the law-court is a strange place.[11] As Johnstone put it, following the work of legal anthropologist Sally Falk Moore, legal rhetoric in Athens existed in a "semi-autonomous field."[12] It both participated in the value systems and patterns of thought that occurred within wider society and, at the same time, reserved for itself particular modes of operation that were particular to itself.

The law-court constructed particular regimes of truth in which, through the application of certain rules and practices, justice could be delivered. It operated with its own particular logic. Wohl, for example, has focused on the apparent "kettle logic" of a number of legal speeches in which implicitly incompatible arguments are thrown together.[13] As a result, it was possible for a litigant who was charged with being a member of the cavalry during the oppressive regime of the Thirty to argue,

> virtually in a single breath, that he was not in the cavalry nor was his name on the registry, but even if he was on

11 For a discussion of the "discursive specificity of the law-court," see Wohl 2010: 21–37 with bibliography.

12 Johnstone 1999: 121. For the work of Moore, the key text is Moore 1973.

13 The phrase, borrowed from Freud, refers to the joke that "A. borrowed a copper kettle from B. and after he had returned it was sued by B. because the kettle now had a big hole in it which made it unusable. His defense was, 'First, I never borrowed a kettle from B. at all; secondly, the kettle had a hole in it already when I got it from him; and thirdly, I gave him back his kettle undamaged.'" Wohl 2010: 8–9.

> the registry, you can't trust that because the list can be forged; if he had been in the cavalry he would admit it and be proud of his service (which he would prove was spotless); anyway, many other people who were in the cavalry now serve on the Council, and he wouldn't even bother to defend himself on this score if his opponents weren't telling bare-faced lies about him.[14]

The list of peculiarities of forensic rhetoric could be extended. Other examples include its fetishization of arguments based on probability, the preeminence it accorded to evidence of good character, its complicated rules regarding witness statements, or its seemingly paradoxical stance on slave evidence.[15] The logic of this last aspect is particularly telling. According to Athenian legal practice, evidence from slaves was not admissible unless it had been first obtained by torture. However, when it was so obtained, the purifying nature of torture made the words of the slave into the "best evidence possible."[16] Here a belief in the transformative power of pain and the appetitive nature of the slave body gave rise to a particular regime for the production of truth.[17] Moreover, this regime seems to have been exclusively the province of the legal procedure. Elsewhere authors seem all too conscious of the fact that the tortured say just what the inquisitor wants to hear.[18] Indeed, in one case where a slave is tortured outside the law-court, his statements are subsequently rendered inadmissible because they had been contaminated by improper procedure. In such circumstances, torture does not seem to work – "they will say

14 Wohl 2010: 9 paraphrasing Lysias 16.6–8.

15 For the importance of arguments based on probability, see Arist., *Rhet*. 1402a1 and *Rhet ad Alex*. 1442a23–37 and 1443b35–40. For the importance of character, see Arist., *Rhet*. 1377b20–1378a3 and *Rhet ad Alex*. 1441b35–1442a5. For an example in action, see Lys. 21 in which a large portion of the defence simply consists of a recitation of civic benefactions.

16 Dem. 30.37. Cf. Dubois 1991: 49–50.

17 DuBois 1991. Cf. Gagarin 1996 and Mirhardy 1996.

18 See Arist. *Rhet*. 1377a1–5.

anything to gratify their torturers," complains the litigant.[19] Only within the confines of court proceedings will torture work its "magic."

While the peculiarity of the law-court has been observed, much less attention has been placed on the way in which this peculiarity was established within public space. Forensic space didn't just happen; it was made. A lot of effort was required to create an atmosphere in which these particular rules of language, argument, and thought felt acceptable and natural. The rituals and physical environment of the law-court helped to locate it as a space that functioned "in the city" and "as the city." Its physical environment both stressed its separateness from the world and its intimate connection with that world. When one entered the law-court as either juror, litigant, or spectator one knew that certain special rules of physics would apply, but that those rules were not entirely alien.

PERFORMING THE *DIKASTERION* INTO BEING

In a passage preserved in Athenaeus, the fourth-century comic poet Euboulos offers an overblown description of the *kleroterion*, the allotment machine that determined eligibility for jury service and assigned jurors to law-courts. Part joke, part surrealist-fantasy, the description presents us with a challenging idea of what it means to be a juror in Athens:

> ἔστιν ἄγαλμα βεβηκὸς ἄνω, τὰ κάτω δὲ κεχηνός,
> εἰς πόδας ἐκ κεφαλῆς τετρημένον, ὀξὺ διαπρό,
> ἀνθρώπους τίκτον κατὰ τὴν πυγὴν ἕν' ἕκαστον·
> ὧν οἱ μὲν μοίρας ἔλαχον βίου, οἱ δὲ πλανῶνται . . .

> There is a sculpture that is closed on top but wide open at the bottom, drilled sharply from head to foot. It gives birth to men from its bottom, one by one, and some gain the right to live while others are left to wander away . . . (Ath. 10.450b4–8=Euboulos [*PCG* V] F106.21–25)

[19] Antiphon 5.31–32.

Here Euboulos reformulates the allotment machine into a quasi-mystical object, the language elevates a routine administrative practice into a form of religious rite. Central to this riddling description is the idea that being selected to serve as a juror constitutes a form of rebirth. One only really "comes to life" when one is participating in the civic life of the city; if you are not selected to be a juror, you wander away unborn.[20] Furthermore, as the phrase "giving birth from its bottom" (τίκτον κατὰ τὴν πυγήν) signals, there is something slightly unnatural about this rebirth. It is an artificial experience.

Prior to the identification of the marble *kleroteria* in the Agora in 1937, few people understood this passage in Athenaeus.[21] Earlier scholars imagined the *kleroteria* as physical sorting rooms, part of a large complex of buildings. At their most elaborate, they imagined dozens of sorting rooms (two per tribe) providing an entranceway to courtrooms beyond.[22] Given that the -τήριον ending can indicate a structure as well as an instrument, such a misconception is understandable. Certainly the term confused a couple of later antique commentators who made exactly the same mistake.[23]

Yet the misunderstanding about the *kleroteria* is symptomatic of a much wider misconception about the Athenian law-

20 The image is intensified if we include the subsequent line from Athenaeus, often rejected by editors as spurious, which envisages the rejected juror wandering away like a murdered spirit calling out a warning to all it encounters – cf. Dow 1939: 12. There is also another level of allusion. In imagining the allotment procedure giving "life" to the juryman, Euboulos also plays with the idea of the jury pay as providing a "livelihood" to the juror.

21 Although fragments of *kleroteria* had been published as early as 1861 by Stephanos Koumanoudes (*Ἀρχαιολογικὴ Ἐφημερίς* 1862: 25), it was not until Dow 1937: 198–215 that the purpose of these machines was correctly identified. Subsequent discussion of the machines and their use is provided by Dow 1939.

22 See, for example, the schema offered in Hommel 1927: 141. Cf. Sandys' 1893 commentary on allotment procedure in his notes on *Ath. Pol.* 64.1–5. The same arrangement informs Rackham's 1935 Loeb translation of *Ath. Pol.* 64.1–5.

23 Phrynichos s.v. "κληρωτήρια" (Bekker, *Anecdota* 47.13) and Pollux 9.44.

court. Underlying most of the attempts to locate the Athenian law-court during the late nineteenth and most of the twentieth century was the assumption that the *dikasterion* was an impressive single-use structure, a structure not unlike the law-courts that dominated the civic spaces of the hometowns of most of the excavators and topographers.[24] Given the importance that the law-court played in the democratic life of the city, it was felt that only a suitably monumental structure could serve as the city's law-court. It was for this reason that the impressive square structure in the Agora by the South Stoa was identified for a long time as a potential law-court. For the excavators, both the position ("exceedingly prominent") and the quality of the building work ("outstandingly fine") seemed to confirm the location of the "largest and most famous" law-court in Athens, the Heliaia.[25] Neither the complete absence of any dikastic material found in excavation within the building nor its impracticality as a meeting place for the requisite number of jurors seem to have dissuaded them from the identification.[26]

The subsequent reasonably secure identification of the so-called "Heliaia" as the "Aiakeion," a fourth-century granary, confirms the proposition that monumentality is a poor criterion for the identification of forensic space.[27] Law-courts did not need to be housed in large, purpose-built structures. As we shall see, law-courts occupied multipurpose structures. They could be grand, impressive buildings, but they could also be

24 For a discussion of the impact of the rise of the nineteenth-century monumental law-court on preconceptions about the Athenian law-court, see Blanshard 2004a: 13–17.

25 Thompson and Wycherley 1972: 62–65.

26 For distribution of dikastic material in Agora excavations, see Boegehold et al. 1995: fig. 4. For criticisms of the identification of the building as a law-court on practical grounds, see Hansen 1981–82, 1989: 232–35.

27 For the identification of the Aiakeion, see Stroud 1998: 85–104, esp. 94–95. Complicating the story of the identification is the allegation made by A. N. Oikonomides that Wycherley and others involved in the Agora excavation deliberately suppressed evidence (such as POxy 2087) that challenged their own identifications, cf. Oikonomides 1990 with Stroud 1994: 1 n.2.

marginal and poorly built. In this, the pattern of use replicates early modern legal proceedings where law-courts could be constituted in meeting halls, dining rooms, theatres, schools, and public houses.[28] Courts were often makeshift, temporary structures.

A survey of the names for law-courts supports this idea of a shifting, variously constituted legal landscape. At least eighteen different names for the various courts in Athens survive. While in some cases we may have multiple names for the same court, the lack of consistent nomenclature supports the idea of the law-courts being constituted in a variety of locations and taking a variety of forms. We have courts named after people (Kallion, Metiocheion), places (Odeion, Painted Stoa, "against the walls," "greater," "triangular"), colours ("the red court," "the frog-green"), officials ("of the Eleven," "of the Archon"), and age ("new").[29]

Literature helps identify a number of known sites as locations for law-courts. The reference to the law-court at the Odeion in Aristophanes' *Wasps* 1109 must refer to the Odeion of Pericles.[30] There is some evidence in the later commentators that this site was particularly associated with lawsuits administered by the Eponymous Archon.[31] However, this may just be supposition on the part of the commentators based on the fact that the only surviving reference in the Attic orators relates to one such suit.[32] Similarly, inscriptional evidence relating to the activities of the Delphian Amphiktyones helps identify the Stoa Poikile as the site of arbitrations and legal proceedings in the fourth century.[33] The relatively large jury number recorded

28 For discussion of the early modern legal landscape in Europe and the UK, see Graham 2003: esp. 9–34. For similar arrangements obtaining in the US, see Johnson and Andrist 1977 and Waddell 1981 with Blanshard 2004a: 17–19.

29 For a survey of the terms, see Boegehold et al. 1995: 91–98.

30 Boegehold et al. 1995: 185–86 for testimonia.

31 Photios and repeated in the Suda *s.v.* Ὠδεῖον.

32 [Dem.] 59.52, 54.

33 *IG* II^2 1641.25–33. Boegehold et al. 1995: 161–62 provides discussion of the testimonia. For arbitration procedures occurring in the stoa, see Dem. 45.17.

in the inscription helps support the idea that the Stoa Poikile was a suitable site for both private actions and more important public actions with their consequently larger juries.[34]

Less conclusive, but nevertheless persuasive, evidence has been adduced for the benches set on the east slope of Kolonos Agoraios (the "poros benches") as the location of a law-court.[35] The argument turns on a pun in Aristophanes *Wasps* where there is an allusion to a law-court and the name "Lykos" seems to be substituted for the epithet "Lykeios."[36] The pun works most effectively if the audience were expecting a reference to a law-court near a shrine to Apollo (Lykeios). Given their proximity to the temenos of Apollo Patroos and their suitability as a venue, the poros benches would seem to be the most likely candidate for the intended law-court.

Archaeological evidence also helps secure the location of law-courts in some buildings. The most famous example is the "ballot deposit" found in Building A underneath the Stoa of Attalos. Here the use of the structure as a law-court was confirmed by the discovery in 1953 of a collection of small finds relating to the law-court (five bronze voting ballots, a bronze token, and a bronze ball).[37] Surrounding the building there was also a high density of other dikastic material including another three voting ballots and thirteen more tokens. To judge from the small finds, Building A, which was constructed at either the end of the fifth century or the first decade of the fourth century, seems to be part of a complex of buildings (so-called Buildings B-D) which served diakastic functions.[38] At some point towards the end of the fourth century, this complex was demolished and plans were made to replace them with a much

34 *IG* II^2 1641 records a jury size of 499. For discussion of the size, see Boegehold 1984: 28–29 and Stumpf 1987: 211–13. On Athenian jury size varying according to the type of legal procedure and importance of the case, see Harrison 1998: ii.47.

35 Boegehold 1967.

36 *Wasps* 389. For discussion, see Boegehold et al. 1995: 95, 188.

37 Thompson 1954: 58–61. Further discussion is provided in Thompson and Wycherley 1972: 56–57 and Boegehold et al. 1995: 53–54, 68.

38 Boegehold et al. 1995: 11–15.

larger single structure, the so-called Square Peristyle. The design of the building and the presence of letters on a number of blocks indicate that the building was intended to be used as a law-court in a manner consistent with the procedure outlined in the *Athenaion Politeia*.[39]

We can also locate a number of courts geographically even if we can't associate them with particular structures. This is particularly the case with a number of homicide courts. For example, the court of the Areopagus, which dealt with cases of premeditated homicide as well as other religious cases such as violation of sacred property met on the Areopagus, but the complicated archaeology of the site precludes us from identifying with certainty any particular structure as the location of the law-court.[40] The location of the other homicide courts, the Delphinion (for cases of justified homicide) and Palladion (for cases of accidental homicide), is even less precise, although in both cases the general area is known.[41]

A survey of the sites that we can identify either archaeologically or through literary evidence demonstrates the difficulty of generalizing about the physical nature of the law-court. Indeed, this lack of consistency is perhaps the most striking feature of the Athenian law-court. The law-courts vary enormously in size, shape, and floor plan. Some buildings, such as those under the Stoa of Attalos, the Odeion of Pericles, and the Stoa Poikile, were roofed. Others, such as the homicide courts, the square peristyle, and the *poros* benches, were open to the air.[42] To judge from the architectural remains, some buildings were elaborate, highly decorated structures. Others used shoddy,

39 Boegehold et al. 1995: 108–11.

40 For the legal competency of the Areopagus, see MacDowell 1963 and Sealey 1983. Of the surviving forensic orations, Lysias 7 is the only certain speech to have been delivered before the Areopagus.

41 For the location of the Delphinion, see Travlos 1971: 83. A site near Phaleron rather than near the Ardettos hill is probably to be preferred as the location of the Palladion, see discussion in Boegehold et al. 1995: 47–48.

42 For the homicide courts being open to the air, see Antiphon 5.10–11.

secondhand materials. As one report on Athenian civic architecture concludes, "the architectural setting of the Athenian *dikasteria* was indeed inconspicuous and as far removed from monumental architecture as anything can be."[43]

The peculiarity of forensic space is brought into relief when one contrasts the performance space of forensic rhetoric with the two other standard genres of civic rhetoric, the symboleutic and the epitaphic. Rather than a diversity of locations, both are tied to singular locations, the Pnyx and the demosion sema. Both are impressive sites. Prior to the reversal of the Pnyx, the speaker had a spectacular civic vista dominated by the Acropolis as his backdrop.[44] When rhetors spoke about the benefits for and dangers to the city, their audience could see precisely what was at stake. Similarly, the preliminary archaeological excavations of the site of the demosion sema seem to support a suitably impressive site for the delivery of orations for the war dead.

Forensic space, in sharp contrast, relies less on specific architectural forms than on the demarcation of space through procedure. As the fragment of Euboulus that began this discussion suggests, a consideration of the life of the juror must begin with the strange procedure that "gives birth" to him. Picking up on the imagery of mystery cult with its idea of souls waiting to be incarnated, Euboulus' riddle locates the creation of the jury within the overdetermined rituals that accompanied the installation of law-courts and the transformative power of the paraphernalia that signified its presence. Great emphasis was placed on the presence of court paraphernalia. These are the props underpinning the performances that allowed the stoa, Odeion, square peristyle or open-air bench to become a *dikasterion*. It was the temporary and movable fences (*druphaktoi*) and lattice-work gates (*kinklides*) that helped make a law-court.[45]

43 Hansen and Fischer-Hansen 1994: 77.

44 Hansen 1982 and 1985 provide a discussion of the topography of the Pnyx and the history of its phases. For the rhetorical use of this vista, see Aeschin. 2.74.

45 For the light, temporary railings and swinging gates used to mark out the limits of the law-court and control the movement of

Another work by Euboulos picks up on this aspect. In a fragment of his *Olbia*, Euboulos has two characters present differing perceptions of the nature of Athens:

ἐν τῷ γὰρ αὐτῷ πάνθ' ὁμοῦ πωλήσεται ἐν ταῖς Ἀθήναις·
A: σῦκα
B.: κλητῆρες
A.: βότρυς, γογγυλίδες, ἄπιοι, μῆλα
B.: μάρτυρες
A.: ῥόδα, μέσπιλα, χόρια, σχαδόνες, ἐρέβινθοι
B.: δίκαι
A.: πυός, πυριάτη, μύρτα
B.: κληρωτήρια
A.: ὑάκινθος, ἄρνες
B.: κλεψύδραι, νόμοι, γραφαί

You will find everything sold together in the same place in Athens:
A: Figs.
B: Issuers of summons.
A: Grapes, turnips, pears, apples
B: Witnesses
A: Roses, medlars, caul, honeycombs, chickpeas
B: Private legal actions
A: Colostrum, cholostrum cheese, myrtle
B: Allotment machines
A: Hyacinth, lambs
B: Water-clocks, laws, and public indictments (Ath. 14.640b = Euboulus [*PCG* V] F74)[46]

From the beginning of the fragment, we know that the topic for discussion is going to be location. The fragment grounds its humour in the juxtaposition of two different worlds – the *emporion* and the *dikasterion*. However, the fragment is not overly concerned with these worlds, rather its primary focus

people entering and exiting forensic space, see Boegehold et al. 1995: 195–201.

46 The decision to split the fragment between two voices followed by Kassel-Austin was first suggested by Toeppel in 1851.

is another place – Athens. The city provides the matrix that structures the relationship between these two *topoi*.

This focus on location is also highlighted by the context of this quotation. The fragment occurs in a discussion on edible delicacies and is introduced with the phrase ὡς Εὔβουλος δ'ἐν Ὀλβίᾳ ἔφη ("As Euboulos says in *Olbia*"). Immediately, we are confronted with a question – How are we to read ἐν Ὀλβίᾳ? Three possibilities present themselves.

Firstly, it could refer to a play about an imaginary location. It "recalls the tradition of the comic *Schlaraffenland* as it is found in a number of passages from Old Comedy preserved by Athenaeus."[47] In such a scenario, the poet's literary construction of Athens takes on vestiges of reality through comparison with an unattainable ideal. Secondly, Olbia might not be a place, but a literary *persona*. The context of the fragment might be a *hetaira* play.[48] However, even here place is never far from the surface. For the *hetaira* play is a genre driven and structured by the fact (anxiety/relief) that the *hetaira* must always be outside the *polis*.[49] Thirdly, it could refer to a physical location; Olbia on the northern shore of the Black Sea. On this reading, we should envisage only one voice – the poet's.[50] Our fragment is the comic's explanation of his home to a foreign audience. This is how Athens presents itself to (admiring/threatening/indifferent . . .) others. Wherever we turn, this fragment confronts us with the concepts of place, boundaries, and division.

This concern with boundaries is brought out in the content of the fragment. The fragment plays with the transparency of the law-court. The additions of "κλητῆρες . . . μάρτυρες . . . δίκαι . . . κληρωτήρια . . . κλεψύδραι, νόμοι, γραφαί!" is both a continuation (metrical and conceptual) of the dialogue and a sharp, discontinuous interjection. Although part of the

47 Hunter 1983: 164.

48 Kaibel *RE* VI 1 1907: 878 followed by Edmonds 1957–61: ii.115.

49 For the complexities of place, topography, and the Athenian discourse of the courtesan, see Gilhuly's essay (Chapter 5) in this volume.

50 See Hunter 1983: 164.

humour of this passage depends on the fact that the majority of law-courts were located either in or near the Agora, this interchange is not just about physical proximity. The dialogue asks us to theorize the relationship between the marketplace and the law-court. What does it mean to have the marketplace by the law-court? How great is the distance (physical, ideological) between them? To what extent is the law-court like the marketplace? To what extent is it so unlike the marketplace that its presence in the Agora inevitably colors the marketplace, and marks it as unusual (i.e., Athenian)?[51]

The poet raises these concerns by defining the law-court in terms of commodities. The *dikasterion* is delimited by a series of movable or unfixed objects. Like the marketplace, the law-court is expressed through the objects that it displays. The target is well-chosen. The law-court was particularly susceptible to such techniques of representation. Law-courts were signified by a large amount of specialist equipment. For example, in his lexicon of rhetorical terms, Pollux lists over 18 separate items that are found in *dikasteria* (e.g., *klepsydrai*, hoppers, urns, trays, staves, painted boards).[52]

The "court scene" in Aristophanes *Wasps* attests to the semiotic effect of court paraphernalia.[53] Through its comic substitution of domestic goods for various features of the law-court, culminating with Philocleon turning his chamber pot into a water clock, this scene plays on the ease with which domestic space can be translated into forensic space. This scene plays humorously with the idea of the collapse of public and private spaces.[54] The joke here is double-pronged. It is not just that the

51 Such collocations, for example, were particularly troubling to Aristotle, who advocated the establishment of two *agorai* – one free (*agora eleuthera*) and one necessary (*anagkaia agora*) -Arist. *Pol.* 1331a30-b14, cf. Millett 1998: 218–20.

52 Poll. 8.16–18. For the collection of testimonia relating to the minor movable court equipment, see Boegehold et al. 1995: 208–41.

53 *Wasps* 805–1008.

54 On public and private space as one of the central themes in the *Wasps* (e.g., from public *dikasterion* to private *symposium*), see Crane 1997.

court shouldn't easily/readily be in a private house, it is also that it *can* easily/readily be in a house.[55]

It was through the ritualized use of such paraphernalia that a sense of space and occasion was created. Indeed, there was so much focus on procedure and equipment that Athenian court-life seems, at times, remarkably overdetermined.[56] For example, it was not just sufficient that there be a special water clock to time the length of legal speeches, the water for the clock needed to be collected from a special lion-fountain and the bung administered by a ceremonial bronze hammer.[57]

Just as forensic rhetoric grows in importance throughout the course of the fourth century BC, so too does the ritual attending to the creation of forensic space.[58] What begins in the fifth century as a relatively simple method of jury selection and voting procedure turns into an elaborate performance by the final quarter of the fourth century, which involves dozens of officials, elaborate allotment procedures governing every aspect of decision-making, and great processions of jurors passing through the Agora carrying colored staffs and acorns as a sign of their office. At one point, Demosthenes reminds jurors about the symbolic importance attached to their staves.[59] He is right to do so; the existence of the law-court depends on them.[60]

Staging the Actors and the Audience

Given the intense investment involved in creating forensic space, it is worthwhile investigating what effect this spatial construction had on participants in legal actions. The nature of

55 Blanshard 2004a: 23.

56 "Why the rigmarole?" Bers 2000 asks in the title of an article on allotment procedure. For discussion of the overelaborate nature of proceedings, see Bers 2000: 554–56.

57 Pollux 8.113 (lion-fountain) and 10.61 (hammer).

58 On the increasing power and politicization of the law-court throughout the fourth century, see Hansen 1974 and Todd 1993: 156–63.

59 Dem. 18.210.

60 For the importance of how one moves into, out of, and through interior spaces, see Purves' essay (Chapter 3) in this volume.

space inevitably conditions how people feel, react, and behave. For example, much research has been conducted into understanding how modern litigants experience the courtroom environment and recent studies have shown how the environment influences subjects' patterns of thought and speech.[61] Following the implications of such studies, it is profitable to examine what aspects the court environment stressed and how this affected the role of the various participants in that environment. Following trends in the phenomenological analysis of archaeological space, this analysis opens with an analysis of dikastic experience before moving to a treatment of programmatic aspects of law-court architecture.[62]

For the jury, the creation of forensic space was both potentially empowering and troubling. Ever since Ober (1989), the courtroom has regularly been constructed as a place where the *demos* (as embodied by the jury) and democratic processes are praised and validated.[63] Even when the equation of the *demos* with the jury has been questioned, the implicit democratic nature of the courtroom has remained unassailed.[64]

Such endorsement of democratic values certainly occurred, but it does not follow from this that jury service was an unproblematic pleasure. Aristophanes' *Wasps* portrays the life of the jury as one of jovial bonhomie. They know and support each other. As befits a chorus, they often move and speak as one. Yet, it would be a mistake to translate such solidarity to the world outside of the comic stage.

[61] The literature on the topic is voluminous, although predominately dominated by studies examining the effect of courtroom design on vulnerable participants (e.g., children, the mentally ill). See for example, Karras et al. 2006: esp. chapter 5; and Kennedy and Tait 1999.

[62] For an introduction to the issue of archaeology and the phenomenology of space, see Tilley 2010: esp. chapter 1.

[63] For bibliography, see Wohl 2010: 32 n. 28. For criticisms of this position and the presentation of a modified version, see Wohl 2010: 32–37.

[64] For debates about the relationship between the demos and the jury, see Blanshard 2004b with bibliography. The most recent contribution to this ongoing debate is Hansen 2010.

Whatever value Aristophanes' depiction has for the fifth century, a very different situation obtains for most of the fourth century.[65] The fourth-century juror may have arrived with friends to undertake jury service, but he was soon parted from them. The city did not accept every juror who appeared for service. Instead, as we have seen above, they had to pass a random selection procedure. Firstly, at some point before he began his service, a juror's allotment ticket was randomly marked with a letter between A and K. This performed an initial separation amongst the jury pool as jurors with tickets marked A and E were allocated separately from jurors whose tickets were marked with a letter between I and K. Even if it were the case that the juror's associates were allocated to the same half of the alphabetic division, their chances of being allocated to the same court were slim. Jurors were not allowed to load the allotment machine themselves, this duty was performed by a member of the tribe chosen by sortition precisely to prevent the possibility that jurors could get themselves allocated to the same potential pool.[66] In addition, the member charged with loading the machine inserted the tickets into the machine randomly so that even if he was familiar with the other jury members it was impossible for him to allocate friends and associates together. Finally, the operation of the machine itself randomly selected and rejected panels of jurors according to whether a white or black ball was drawn at the point at which the panel was being considered for jury duty.

As one can see, the chances of being successfully selected to serve as a juror with one's friends were slim. However, even if it were the case that you knew people within the pool of successful jurors, the city undertook two further forms of sortition in order to break up such social groups. Firstly, each juror was not allowed to select which court they sat in. This was determined by sortition.[67] Secondly, within the courtroom,

[65] For changes in the procedure for the allotment of jurors between the fifth and fourth centuries, see Harrison 1998: ii.239–41.

[66] *Ath. Pol.* 64.2.

[67] *Ath. Pol.* 64.4.

jurors were not allowed to choose their seating; this also was randomly determined.[68]

It is worth stressing the spatial isolation of the juryman in the law-court because it marks an almost unique occurrence in the city – a moment when a citizen is forced to make a decision not surrounded by family and friends.[69] Every other major life decision – whether it involved marriage, business, family, or politics – was always done in the company of associates.[70] No such procedures for isolating citizens occurred in the *ekklesia*. Indeed, all available evidence supports the contrary view, namely that people sat in blocks in the Assembly. In the law-court, for perhaps the first time, the juror found himself alone.

In such a circumstance, little comfort was provided by the orators who continually warned jurors about the deceptive power of rhetoric.[71] Speakers further ratcheted up the stakes by reminding jurors that not only did their decisions have consequences for the security of the state, but that also their verdicts were subject to close scrutiny. The voting procedure may have been anonymous, but the orators refuse to allow this to be any comfort to jurors. In *Against Aristogeiton*, the speaker reminds each juror how the goddesses Eunomia and Dike "who sits beside the throne of Zeus" watch over the juror as he casts his vote making sure that he doesn't dishonour the Gods.[72]

68 *Ath. Pol.* 65.2 with Boegehold 1984: 23–29.

69 The allocation of random seating in the Bouleuterion provides a parallel (see Philochoros *FGrHist* 328 F140) although here the sense of isolation was presumably lessened by the continuity of service among a stable group.

70 See, for example, Lysias 32.12–18 (family gathered to discuss inheritance and commencement of legal proceedings), Dem. 54.1 (on consultation with friends and relatives over legal action), Isaeus 2 (on the discussion with friends over issues of marriage and adoption). The *hetairiai* of Thucydides (8.48 and 54) provide perhaps the most famous example of friendship groups influencing political decisions, see Calhoun 1913 and Sartori 1967.

71 On rhetoric as a form of magic (*goeteia*) see Din. 1.66, Aeschin. 3.137, 207. Cf. de Romilly 1975.

72 [Dem.] 25.10–11.

Apollodorus, on the other hand, relies not on supernatural forces, but on more prosaic domestic arrangements to remind the juror that there are consequences if he gets his decision wrong:

> τί δὲ καὶ φήσειεν ἂν ὑμῶν ἕκαστος εἰσιὼν πρὸς τὴν ἑαυτοῦ γυναῖκ' ἢ θυγατέρα ἢ μητέρα, ἀποψηφισάμενος ταύτης, ἐπειδὰν ἔρηται ὑμᾶς 'ποῦ ἦτε;' καὶ εἴπητε ὅτι 'ἐδικάζομεν;' 'τῷ;' ἐρήσεται εὐθύς. 'Νεαίρᾳ' δῆλον ὅτι φήσετε (οὐ γάρ;) 'ὅτι ξένη οὖσα ἀστῷ συνοικεῖ παρὰ τὸν νόμον, καὶ ὅτι τὴν θυγατέρα μεμοιχευμένην ἐξέδωκεν Θεογένει τῷ βασιλεύσαντι, καὶ αὕτη ἔθυσε τὰ ἱερὰ τὰ ἄρρητα ὑπὲρ τῆς πόλεως καὶ τῷ Διονύσῳ γυνὴ ἐδόθη', καὶ τἆλλα διηγούμενοι τὴν κατηγορίαν αὐτῆς, ὡς καὶ μνημονικῶς καὶ ἐπιμελῶς περὶ ἑκάστου κατηγορήθη. αἱ δὲ ἀκούσασαι ἐρήσονται 'τί οὖν ἐποιήσατε;' ὑμεῖς δὲ φήσετε 'ἀπεψηφίσμεθα.' οὐκοῦν ἤδη αἱ μὲν σωφρονέσταται τῶν γυναικῶν ὀργισθήσονται ὑμῖν, διότι ὁμοίως αὐταῖς ταύτην κατηξιοῦτε μετέχειν τῶν τῆς πόλεως καὶ τῶν ἱερῶν.
>
> And so what are you going to say when you get home, having acquitted this woman, when your wife or daughter or mother asks, "so where were you?" and you answer "doing jury service." "Judging whom?" will immediately be the reply. "Neaira," you will say (won't you?) "that woman who although a foreigner has been living with a citizen against the laws and who gave her slatternly daughter in marriage to Theogenes, the Basileus Archon, and whose daughter performed solemn rites on behalf of the city and was given as a symbolic bride to Dionysus." And you give a full account of all the charges against her, showing how each charge was well made out. And so hearing this, the women will ask "What did you do?" and you will reply "We acquitted." At this the most chaste of the women will rise up in anger against you because you have judged this woman to have a share with them in the affairs and rites of the city. ([Dem.] 59.110–11)

The passage begins with an injunction to the jurors that they will be called to account for their decision of acquittal. This injunction, with a slight anacoluthon, transforms into a hypothetical dialogue between a juror and a disembodied (Is it wife, daughter, mother, or other?) interrogator. The passage asks the jurors to envisage a domestic scene – but what type of domestic scene? This passage has often been taken by scholars to exemplify a typical interchange between a man and a woman (often taken as a man and wife).[73] However, such a reading ignores the hypothetical and literary nature of this dialogue. This is not a dialogue from real life, but one from anxieties and fears. This is the stuff of nightmare. It is a world where women are the interrogators and, in a dramatic reversal, men are found accountable to them. It is a place where those who were deemed ideologically, legally, and physiognomically incapable of being jurors are now judging those who are above them. It is a world where the juror cannot avoid the questions or fudge the issues but is forced to condemn himself through his litany of Neaira's crimes. He is allowed no cloak of rhetorical tricks to justify his decision. He knows that he has reached the wrong verdict and yet is forced to confess to the one constituency that he knows will take most offence. Like all nightmares, time runs askew and the scene fades out to women in riot. The righteous are zealously fired up by justified anger while the foolish cavort in their reckless decadence.

The jurors found themselves in a far from enviable position. Validation of democratic procedures certainly occurred, but not without effort and anxiety on the part of those involved. The ritualized construction of space served to alienate the juror from the world in which he was most comfortable. The juror never forgot that he was in the city, but that city – its networks of familial and social relations – certainly seemed distant from him as he made his decision. It watched and judged him, but was unable to help him. Forensic space turned out to be a lot more dangerous than first envisaged. Confronted by a version of the "Cretan liars" paradox in which it was impossible to tell who was lying and who was telling the truth, the only

[73] It is perhaps significant, for example, that this passage never seems to be cited for descriptions of father-daughter relations.

certainty that the juror faced was that his reentry from the "forensic" to the "real" world could well be bumpy.

In addition to a phenomenological reading of forensic space, we might also examine how the physical environment of the law-court acted as a generative space. In this respect, it is worth considering which issues the court environment foregrounds and the potential for thematic synergy between word and place. Orators were certainly aware of the potential of architectural exegesis. We get a sense of the potential richness of the interaction between word and place in Aeschines, *Against Ctesiphon*. As part of his attack on the proposal to award honours to Demosthenes, Aeschines draws inspiration from the monuments of Athens. They do not support the notion of the awarding of conspicuous honours to individuals:

> Προέλθετε δὴ τῇ διανοίᾳ καὶ εἰς τὴν στοὰν τὴν ποικίλην· ἁπάντων γὰρ ὑμῖν τῶν καλῶν ἔργων τὰ ὑπομνήματα ἐν τῇ ἀγορᾷ ἀνάκειται. Τί οὖν ἐστιν, ὦ ἄνδρες Ἀθηναῖοι, ὃ ἐγὼ λέγω; ἐνταῦθα ἡ ἐν Μαραθῶνι μάχη γέγραπται. Τίς οὖν ἦν ὁ στρατηγός; οὑτωσὶ μὲν ἐρωτηθέντες ἅπαντες ἀποκρίναισθε ἂν ὅτι Μιλτιάδης, ἐκεῖ δὲ οὐκ ἐπιγέγραπται. Πῶς; οὐκ ᾔτησε ταύτην τὴν δωρεάν; ᾔτησεν, ἀλλ' ὁ δῆμος οὐκ ἔδωκεν, ἀλλ' ἀντὶ τοῦ ὀνόματος συνεχώρησεν αὐτῷ πρώτῳ γραφῆναι παρακαλοῦντι τοὺς στρατιώτας.

> Contemplate then the Stoa Poikile – for the memorials of all our finest moments are set up in the Agora. What then am I speaking about, Citizens? The battle of Marathon painted there. Who is the general? Asked this, you would all respond "Miltiades." Any yet, this is not written there. Why? Did he not ask for this boon? He asked, but the people did not give it, but instead of his name, they proposed that he should be painted at the front, urging on his men. (Aeschin. 3.186)

The monument teaches the jury a moral lesson about status, individuals, and the collective.[74] It is made to dramatize the

74 It is not only Aeschines who uses the Painted Stoa for ethical instruction. Similarly, Demosthenes draws a moral lesson from the painting

tension between mass and elite. Aeschines takes the jury to the Stoa Poikile in their imagination. What is striking is that, as we have seen, for some juries having the battle of Marathon before them wasn't just a rhetorical fiction, but a physical reality.

The juries who sat in the Stoa Poikile found themselves confronted by a collection of images whose associations provided a dense background for forensic rhetoric. In addition to the painting of Marathon, there were also painted panels depicting the battle between Athenian and Spartan at Oinoe, the repulsion of the Amazons by Theseus, and the capture of Troy.[75] The Stoa may also have contained a depiction of the supplication of the Heraclidae.[76] These images were complemented by an altar, as well as a display of a number of enemy shields captured by Athenian forces.[77] Such a collocation of material must certainly have complicated the rhetoric that the jury experienced. Speakers' claims to masculinity, patriotism, and self-worth were measured before the massed ranks of the heroes of Marathon. Generals and politicians had their claims weighed against the defeat of Troy and the Amazons. Speakers reminded juries of the importance of justice as they contemplated the rape of Cassandra and its consequences. Litigants begged for mercy before the image of the supplicating Heracleidae.

Time and again, we are presented with such collocations. Trials for justified homicide occurred in a court near the sanctuary of Delphinian Apollo – the meaning of each spoken word triangulated between the intimate proximity of the cult statue of Apollo, the Acropolis visible through the open roof, and

about the eagerness and dependability of the Plateans in assisting their allies (Dem. 59.94). On the influence of the Painted Stoa in Athenian ideology, see Harrison 1972; Wycherley 1953: 29; Francis and Vickers 1985.

75 Following Pausanias 1.15.1–16.1, the problem of the "Oinoe" painting is longstanding. For discussion of the issue, see Taylor 1998: esp. 223–25.

76 Scholiast on Ar. *Pl.* 385. This suggestion was received favourably by Wycherley 1953: 26 who noted the presence of the nearby Altar of Pity, another monument associated with the Heraclidae.

77 D.L. 7.1.14. (altar), Pausanias, 1.15.1 (shields).

the temple of Olympian Zeus nearby.[78] On high the cult center of Athena watched over her institution, the homicide courts. Below litigants find themselves in the hands of Apollo – a god who as the *sunegeros* of the vengeful Orestes is familiar with the courts and their intrigues. He watches approvingly as litigants defend their right to kill the adulterous.[79] He is there always ready to step in for any whose words might falter; ready to offer absolution so that justice might be done.[80] His actions are backed up by Zeus, the deity for whom justice was a special concern.[81]

Alternatively, consider the trials concerning sacred olive trees that occurred on the Areopagus. Here the location and its symbolic importance insured that the question of the oil from the olives could never be treated as just a matter of taxation.[82] Perched on the "Hill of Ares," who could forget what was at stake in these trials. Topography speaks history. Above them towered the Acropolis, whose Erectheion reminded those who sat below that it was the olive tree that bound Athens and Athena together. The olive was a prize worth fighting for. The Acropolis still bore the scars of Poseidon's wrath at his inability to produce anything to rival it. Not even the might of the Persians could destroy it.[83] Symbolic of Athens, it would only flourish stronger and hardier. Athens' destiny and prosperity were entwined in the boughs of its olive trees. In the Areopagus, discussions of olive trees could never just be about guaranteeing an oil-supply.

78 For the open roof of homicide courts, see Ant. 5.11. Indeed, there is some suggestion that the sanctuary may have been unroofed – the open roof seeming to have been associated with Theseus disturbing the workman before they could finish it. A version is preserved in Pausanias 1.19.1.

79 For an example of precisely this collocation, see Lysias 1.

80 On the role of Apollo in the *Eumenides*, see Sidwell 1996.

81 On the justice of Zeus, the classic text is Lloyd-Jones 1971. For the way in which the Persian Wars infused Athenian topography with meaning, see Dougherty's essay (Chapter 4) in this volume.

82 Compare, for example, the prosaic discussion of the issue in *Ath. Pol.* 60.2 with the rhetoric of Lysias 7.

83 Hdt 8.55. Cf. Virgil, Georgics 2.30–31; Pliny, N.H. 17.241.

Sadly, matching specific locations with specific speeches is, in most cases, an impossible task. Such information did not interest the compilers and publishers of the law-court speeches, who predominantly valued them as examples of rhetoric rather than as historical documents. However, that does not mean that we cannot generalize about the themes that dominate the architecture of the sites of the law-courts and the range of associations that they set up.

One of the striking features of the known law-court sites is their strong association with Athenian military (and consequently imperial) power. We have already seen this martial ideology in the paintings that decorated the Stoa Poikile. We find a similar architectural rhetoric in the Odeion of Pericles.[84] Of particular significance is its supposed mimicry of Persian forms (esp. its roof, many columned hall, and lofty south facade). This imitation was a reference to the Odeion's prototype, the tent of Xerxes captured during the Persian wars.[85] The Odeion was an "elaborate victory monument, built of captured booty and using the architectural forms of the defeated enemy for special effect."[86] Even its woodwork was associated with the Athenian victory. Later tradition records that its roof was constructed from the masts and spars of the Persian ships captured after Salamis.[87] Like the Parthenon and Propylaia, this was a building that could be used as a justification for hegemony.[88]

84 The following is indebted to the recent important reading of the Odeion's architecture by Miller 1997: 239–42.

85 Plu. *Per.* 13.9. On the capture of the tent of Xerxes at Plataea, see Hdt. 9.70, 82.1 cf. Paus. 1.20.4.

86 Miller 1997: 239 citing Von Gall 1979: 446.

87 Vitr. 5.9.1. On the structural impossibility of these spars as main beams for such a structure, see Izenour 1992: 30. However, had the wood been maintained, it could have been used as rafters, purlins or battens. For secondary timbers in Greek roofs, see Hodge 1960: 60–75, cf. Meiggs 1982: 471.

88 Isoc. 15.234. See the remarks of Nouhaud 1982: 225: "Les monuments construits au Ve siècle, à l'époque de la splendeur athénienne, encore visibles par les contemporains des orateurs du IVe, offraient à ces derniers des symboles commodes qu'ils n'ont pas manqué d'exploiter . . . Symboles du rayonnement d'Athènes, les monuments sont . . . considérés comme le signe d'une bonne politique."

Its function as a place for the storage of imperial tribute and the arms and armour of the children of the Athenian war dead seems particularly appropriate.

The power and majesty of the city would not have been lost on the audience perched on the poros benches either. The speaker who addressed them found an unrivalled civic panorama for his backdrop. The audience, overlooked by the imposing facade of the Hephaisteion, gazed out over a tree-lined square.[89] In front of them was the Panathenaic Way, a reminder of that most important of civic festivals. The benches were prime vantage spots for another great civic display, the elaborate, ceremonial ride of the cavalry from the Stoa of the Herms to the Eleusinion diagonally opposite.[90] On the far right, stood the Acropolis. At their feet were the stoas and public buildings that housed the organs of government necessary to civic life. On their left was the Altar of the Twelve Gods, the topographical centre of Athens.[91] The use and effect to which such structures could be put is exemplified by Aeschines, who recalls the time that during political debate the speakers put aside questions of caution and safety and just asked the audience "to gaze at the Propylaea of the Acropolis and remember the battle of Salamis and the tombs and trophies of the ancestors."[92]

FORENSIC SPACE AND GENERIC AFFILIATION

Speakers in the law-court not only brushed up against imperialistic building programs, they also ran into other genres of performance. Again, a consideration of the Odeion of Pericles is illustrative. The Odeion of Pericles not only functioned as a

89 Thompson and Wycherley 1972: 21. However, as Millett 1998: 213 notes, this picture ignores the "hundreds (if not thousands) of inscribed *stelai* standing in front of the monuments and buildings."

90 Xen. *Eq. Mag.* 3.2. Cf. Spence 1993: 186.

91 On distances measured from the altar, see Hdt. 2.7; *IG* ii^2. 2640. On the importance of this monument, see Thompson and Wycherley 1972; 129–36; Camp 1986: 40–42; Millett 1998: 211.

92 Aeschin. 2.74.

law-court, it was also intimately connected with the theater. Its physical proximity made it an important feature in the dramatic landscape, and it was home to the *proagon* as well.[93]

The similarity between the *proagon* and a trial creates a nexus of interpretative possibilities. Not only did it involve the selection of judge by lot, but like a trial, the *proagon* was an opportunity of assessing a person's character (in this case the poet) in front of the gaze of the Athenian public. Socrates says he recognized Agathon's *andreia* and *megalophrosune* because of the way in which he conducted himself during his *proagon*.[94] Moreover, given that poets seem to have offered a prose summary of their play, the *proagon* provided an occasion where two otherwise distinct performances took on a similarity of form.[95] In the Odeion, we are confronted with the very real possibility of citizens turning up twice to hear tales of jealousy, love, betrayal, adultery, murder, and families torn apart; one time as theater-goers, the other as jurors.

The relationship between drama and the law-court has been the subject of much scholarship.[96] As Hall concludes:

> Athenian legal speeches reveal affinities with drama in terms of the context in which they were performed, the relationship between speakers and audiences, the enactment of fictive identities even extending to the attention paid to appearance, costume, use of the eyes, gait, and demeanour; and the exploitation of the courtroom, witnesses and other individuals .... Successful performance at a trial required identical skills to those required by

93 For the Odeion as an important spatial referent in drama, see Wiles 1997: 54–56.

94 Pl. *Smp.* 194 a-b.

95 See Scholiast on Aeschin. *Ag. Ctes.* 67.

96 Although one of the earliest studies concluded that "this study ... has, in the matter of comparison, proved less fruitful than I had anticipated, and ... the conclusions are frequently of a negative rather than a positive nature character" (Thomson 1898: vi), subsequent studies have been more successful in establishing the connection between the two genres. For example, see Dorjahn 1927; Perlman 1964; Hall 1995.

> the dramatic actor – stamina, exciting delivery, vocal virtuosity, memorisation, extemporisation, and the abilities to control the audience, hold its attention, and arose its emotion.[97]

Here "context" refers to the idea of aristocratic *agon* conducted in civic space.[98] As we have seen, however, there was also an overlap in physical context. The relationship with drama flowed both ways. Forensic rhetoric was happy to borrow from drama, whether it was to characterize a murdering mother as Clytemnestra or an unwelcome oligarch as Orestes.[99] Similarly, drama was happy to borrow the language and philosophy of the law-court. Both tragedy (*Eumenides*) and comedy (*Wasps*) stage dramas around law-courts, but the relationship goes deeper. The reasoning of the law-court provides the background against which tragedy could explore issues of anger.[100] The entire genre of new comedy would not function without the legal framework that motivates so many plots and character developments.[101] One sees the tension of the intensity of the relationship in the never-quite-comfortable equation of the rhetor with the actor.[102]

Whichever law-court we examine, we encounter similar conflations between genres and spaces. The jurors who sat in the Agora (whether the poros benches, the Stoa Poikile or in Building A-C) adjudicated debates on the meanings of truth, justice, and the limits of knowledge – debates which mimicked

97 Hall 1995: 57.

98 On the isomorphism of the aristocratic *agon* that united dramatic festivals, athletics competitions, meetings of the assembly, and court cases, see Hall 1995: 39 following Garner 1987: 3.

99 Antiphon 1.17 (Clytemnestra); Lys. 13.79 (Orestes) w. Allen 2005: 378–79. Cf. Porter 1997.

100 Allen 2005: 380–86.

101 On the intimate relationship between new comedy and law, see Scafuro 1997; Wallace 2005: 369–72; Lape 2001. For humour in law-court speeches, see Bonner 1922; Cox 1989.

102 The tension gets its most concrete realization in the rhetoric surrounding Aeschines. For discussion of some of the tensions, see Easterling 1999: 159–60.

those conducted by the sophists just outside, who only days before had "held court" in almost exactly the same place as the one where the orator now stood. The lessons that the jury received in history were conducted in the very monuments that stood as evidence to that history.[103] When an orator mentions the atrocities of the Thirty, how must it have affected the jury to know that at that very same spot that regime stationed its cavalry or condemned people to death?[104] What must it have meant to discuss Pericles, Miltiades, or Themistocles in the shadow of the Parthenon, or flanked by the paintings of Marathon or sitting beneath a roof composed of the masts of ships defeated at Salamis?[105]

It seems too much of a coincidence that Attic oratory – this most intertextual of genres, which combines allusions, metaphors, and quotations from history, poetry, mythology, and sculpture – should be located in structures that often combined and housed elements of these various discourses.[106] The multifunction architecture served to conflate generic associations. Rhetoric is a genre that constantly seeks to mask itself and prefers every other name to its own.[107] In the intertextual environment of the law-court, rhetoric may have found itself a place to hide.

CONCLUSION

ξυλλεγέντες γὰρ καθ' ἑσμοὺς ὥσπερ εἰς ἀνθρήνια,
οἱ μὲν ἡμῶν οὗπερ ἄρχων, οἱ δὲ παρὰ τοὺς ἕνδεκα,

103 On the use of history in forensic oratory, see Perlman 1961; Pearson 1941; Nouhaud 1982; Ober 1989: 177–82; Worthington 1992: 18–20.

104 For references to the Thirty, see Nouhaud 1982: 307–16. For example, it is conceivable that the 1,500 killed by the Thirty mentioned by Aeschines (2.77) are the same as the 1,400 sentenced to death in the Stoa Poikile (D.L. 7.1.5).

105 On references in the orators to these individuals, see Nouhaud 1982: 166–69 (Themistocles), 169–77 (Miltiades), and 221–23 (Pericles).

106 On the intertextuality of rhetoric, see, for example, Ford 1999.

107 Hesk 1999 and Wohl 2010: 37–50.

> οἱ δ' ἐν ᾠδείῳ δικάζουσ', οἱ δὲ πρὸς τοῖς τειχίοις
> ξυμβεβυσμένοι πυκνόν, νεύοντες εἰς τὴν γῆν, μόλις
> ὥσπερ οἱ σκώληκες ἐν τοῖς κυττάροις κινούμενοι.

> For we gather together in swarms as if in hives, some of us judging in the Archon's court, some before the Eleven, some in the Odeion, some against the walls, bunched close together, stooping towards the ground and hardly moving like grubs in their cells. (Ar. *Wasps* 1107–11)

In 1897, an Athenian law-court vanished from the Attic landscape. Unlike other acts of archaeological vandalism, this act of wanton destruction was not that of an invading army or overzealous collector of antiquities, but the textual critic's pen. In line 1109 of Aristophanes' *Wasps*, Starkie proposed, contrary to the manuscript tradition, that ὧδε be read for οἱ δὲ in the second clause.[108] He thereby eleminated from our record the presence of the court "against the walls."

This emendation occurs in the context of a metaphor verging out of control. The metaphor comes from the second half of the parabasis. The chorus of jurors had already explained their main point of similarity with wasps; like them, they too have a vicious sting. In this passage, they decide to extend the simile and compare themselves to wasps in terms of lifestyle and habits. They describe themselves as being like wasps with respect to the way in which they cluster together. It is the location of this clustering that is the focus of Starkie's emendation to the text.

Starkie was adamant in his belief that the text was corrupt at this point ("It may, I think, be dogmatically stated that this line cannot allude to a court 'situated near the walls'"). He based his certainty on the fact that there was "no evidence for the existence of such a court." This immediately raises the question, "What sort of additional evidence did Starkie require to confirm the court's existence?" Clearly, Starkie envisages a

108 Starkie 1897: 421. The emendation has been followed in the editions of Henderson 1998 and Wilson 2007.

law-court as a structure that must inevitably leave some sort of trace in either the literary or the archaeological record. He operates under a delusion we have seen disproved above. He imagines the law-court as a structure that has some degree of permanence, exclusivity of use, and specificity of place – in short, a monument.

The significance of Starkie's emendation is that its validity depends almost entirely on our conception of the law-court. Extant manuscripts of the *Wasps* do not exist prior to the eleventh century. There have been no discoveries of text or papyrus applicable to this section. Nor do the normal rules of paleography offer much assistance. The mistranscription of οι for ω (and vice versa) is a relatively common and understandable scribal error. The emendation is not inelegant. It may even be right. However, it is worth unpacking what is at stake in the editorial choice. Our tolerance for a court "hard against the walls" is dependent on the extent to which we are prepared to accept a model of the Athenian court as transient, non-topographically specific, and without consistent structure and form – a court of such (little?) moment that it might be captured for our record only in the verse of a comic poet. A dikasterion that is constituted not as a *topos*, but as a *praxis*.

Starkie's emendation is illustrative of how preconceptions about the physical nature of the courts have an impact on our textual understanding. This chapter has attempted to demonstrate that a fuller appreciation of the materiality of law-court design not only helps us achieve a better understanding of Athenian civic topography, but that it has important consequences for our understanding of forensic rhetoric.

The law-court was the nexus where a number of civic discourses met and where important ideas about citizen identity and the nature of truth could be contested. As such, it proved useful to construct the law-court as a space that stood apart from the rest of civic space. Through the use of ritual and the deployment of specialised paraphernalia it was possible to construct a space with its own idiosyncrasies of language, reasoning, and meaning. Yet, while the space for rhetorical performance stood apart from the city, it never quite severed its attachment to its origins. Place bled into space. The city marshalled the audience and framed the rhetoric. Imperial aspirations, historical contingencies, ideological positions, and the echoes of other literary genres were all present in the places where rhetorical space was formed and they coloured the resultant field of performance. As such, they gave fuel to more expansive readings and complicated the notion of the genre of rhetoric.

Bibliography

Allen, D. 2005. "Greek Tragedy and Law." In *The Cambridge Companion to Ancient Greek Law*, ed. M. Gagarin and D. Cohen. Cambridge. 374–93.

Bers, V. 2000. "Just Rituals: Why the rigmarole of fourth-century Athenian lawcourts?" In *Polis and Politics: Studies in Ancient Greek History*, ed. P. Flensted-Jensen, T. H. Nielsen, and L. Rubinstein. Copenhagen. 553–62.

Blanshard, A. J. L. 2004a. "The Birth of the Law-Court: Putting Ancient and Modern Forensic Rhetoric in Its Place." In *Oratory in Action*, ed. M. Edwards and C. Reid. Manchester. 11–32.

2004b. "What Counts as the *Demos*? Some Notes on the Relationship Between the Jury and 'the People' in Classical Athens." *Phoenix* 58: 28–48.

Boegehold, A. L. 1967. "Philokleon's Court." *Hesperia* 36: 111–20.

1984. "Many Letters: Aristophanes Plutus 1166–67." In *Studies Presented to Sterling Dow on his Eightieth Birthday*, ed. K. J. Rigsby. Durham, NC. 23–29.

Boegehold, A. L. et al. 1995. *The Lawcourts at Athens: Sites, Buildings, Equipment, Procedure, and Testimonia*. Princeton.

Bonner, R. J. 1922. "Wit and Humor in Athenian Courts." *CP* 17: 97–103.

Calhoun, G. M. 1913. *Athenian Clubs in Politics and Litigation*. Austin, TX.

Camp, J. M. 1986. *The Athenian Agora: Excavations in the Heart of Classical Athens*. London.

Cox, C. A. 1989. "Hipponicus' Trapeza: Humor in Andocides 1.130–1." *CQ* 46: 572–75.

Crane, G. 1997. "Oikos and Agora: Mapping the Polis in Aristophanes' *Wasps*." In *The City as Comedy: Society and Representation in Athenian Drama*, ed. G. W. Dobrov. Chapel Hill, NC. 198–229.

de Romilly, J. 1975. *Magic and Rhetoric in Ancient Greece*. Cambridge.

Dorjahn, A. P. 1927 "Poetry in Athenian Courts." CP 32: 85–93.

Dow, S. 1937. *Prytaneis: A Study of the Inscriptions Honoring the Athenian Councillors*. Athens.

1939. "Aristotle, the Kleroteria, and the Courts." *HSCP* 50: 1–34.

DuBois, P. 1991. *Torture and Truth*. New York.

Easterling, P. 1999. "Actors and Voices: Reading Between the Lines in Aeschines and Demosthenes." In *Performance Culture and Athenian Democracy*, ed. S. Goldhill and R. Osborne. Cambridge. 154–66.

Edmonds, J. M. 1957–61. *The Fragments of Attic Comedy*. Leiden.

Ford, A. 1999. "Reading Homer from the Rostrum: Poems and Laws in Aeschines' *Against Timarchus*." In *Performance Culture and Athenian Democracy*, ed. S. Goldhill and R. Osborne. Cambridge. 231–56.

Francis, E. D. and M. Vickers. 1985. "The Oenoe Painting in the Stoa Poikile and Herodotus' Account of Marathon." *Annual of the British School at Athens* 80: 99–113.

Gagarin, M. 1996. "The Torture of Slaves in Athenian Law." *CP* 91: 1–18.

Garner, R. 1987. *Law and Society in Classical Athens*. London.

Graham, C. 2003. *Ordering Law: The Architectural and Social History of the English Law Court to 1914*. Aldershot.

Hall, E. 1995. "Lawcourt Dramas: The Power of Performance in Greek Forensic Oratory." *Bulletin of the Institute of Classical Studies* 40: 39–58.

Hansen, M.H. 1974 *The Sovereignty of the People's Court in Athens in the Fourth Century B.C. and the Public Action against Unconstitutional Proposals*. Odense.

1979. "How Often Did the Athenian *Dicasteria* Meet?" *GRBS* 20: 243–46.

1981–82. "The Athenian Heliaia from Solon to Aristotle." *Classica et Mediaevalia* 33: 9–47.

1982. "The Athenian *Ecclesia* and the Assembly Place on the Pnyx." *GRBS* 23: 241–49.

1985. "Two Notes on the Pnyx." *GRBS* 26: 241–50.

1989. *The Athenian Ecclesia II: A Collection of Articles, 1983–89*. Copenhagen.

1991. *The Athenian Democracy in the Age of Demosthenes: Structure, Principles, and Ideology*. Oxford.

2010. "The Concepts of *Demos, Ekklesia*, and *Dikasterion* in Classical Athens." *GRBS* 50: 499–536.

Hansen, M. H. and T. Fischer-Hansen. 1994. "Monumental Political Architecture in Archaic and Classical Greek Poleis: Evidence and Historical Significance." In *From Political Architecture to Stephanus Byzantius: Sources for the Ancient Greek Polis*, ed. D. Whitehead. Stuttgart. 23–90.

Harrison, A. R. W. 1998. *The Laws of Athens*. Oxford.

Harrison, E. B. 1972. "The South Frieze of the Nike Temple and the Marathon Painting in the Painted Stoa." *AJA* 76: 353–78.

Hesk, J. 1999. "The Rhetoric of Anti-Rhetoric in Athenian Oratory." In *Performance Culture and Athenian Democracy*, ed. S. Goldhill and R. Osborne. Cambridge. 201–30.

Hodge, A. T. 1960. *The Woodwork of Greek Roofs*. Cambridge.

Hommel, H. 1927. *Heliaia. Untersuchung zur Verfassung und Prozeßordnung des athenische Volksgerichs, intsbesondere zum Schlußteil der Ἀθηναίων Πολιτεία des Aristoteles*. Leipzig.

Hunter, R. 1983. *Eubulus: The fragments*. Cambridge.

Izenour, G. C. 1992. *Roofed Theaters of Classical Antiquity*. New Haven, CT.

Johnson, H. A. and R. K. Andrist. 1977. *Historic Courthouses of New York State: 18th and 19th Century Halls of Justice Across the Empire State*. New York.

Johnstone, S. 1999. *Disputes and Democracy: The Consequence of Litigation in Ancient Athens*. Austin, TX.

Karras, M., E. McCarron, A. Gray, and S. Ardasinski. 2006. *On the Edge of Justice: The Legal Needs of People with a Mental Illness*. Sydney.

Kennedy, L. and D. Tait. 1999. *Architecture, Psychology and Western Australian Law Reform*. Perth.

Lape, S. 2001. "Democratic Ideology and the Poetics of Rape in Menandrian Comedy." *CA* 20: 79–119.

Lloyd-Jones, H. 1971. *The Justice of Zeus*. Berkeley.

MacDowell, D. M. 1963. *Athenian Homicide Law in the Age of the Orators*. Manchester.

Meiggs, R. 1982. *Trees and Timber in the Ancient World*. Oxford.

Miller, M. C. 1997. *Athens and Persia in the Fifth Century B.C.: A Study in Cultural Receptivity*. Cambridge.

Millett, P. 1998. "Encounters in the Agora." In *Kosmos: Essays in Order, Conflict, and Community in Classical Athens*, ed. P. Cartledge, P. Millett, and S. von Reden. Cambridge. 203–28.

Mirhardy, D. C. 1996. "Torture and Rhetoric in Athens." *JHS* 116: 119–31.

Moore, S. F. 1973. "Law and Social Change: The Semi-Autonomous Social Field as an Appropriate Subject of Study." *Law and Society Review* 7: 719–46.

Nouhaud, M. 1982. *L'utilisation de l'histoire par les orateurs attiques*. Paris.

Ober, J. 1989. *Mass and Elite in Democratic Athens: Rhetoric, Ideology, and the Power of the People*. Princeton.

Oikonomides, A. N. 1990. "The Aiakeion and Its Relation with the Tholos in the Later Agora." *Ancient World* 21: 21–22.

Pearson, L. 1941. "Historical Allusions in the Attic Orators." *CP* 36: 209–29.

Perlman, S. 1961. "The Historical Example, Its Use and Importance as Political Propaganda in the Attic Orators." *Scripta Hierosolymitana* 7: 150–66.

1964. "Quotations from Poetry in Attic Orators of the Fourth Century B.C." *AJPh* 85: 155–72.

Porter, J. R. 1997. "Adultery by the Book: Lysias 1 (On the Murder of Eratosthenes) and Comic Diegesis." *Echos du monde classique* 16: 421–53.

Sandys, J. E. 1893. *Aristotle. Constitution of Athens: A Revised Text with an Introduction, Critical and Explanatory Notes, Testimonia and Indices*. London.

Sartori, F. 1967. *Le eterie nella vita politica ateniese del VI e V secolo A.C.* Rome.

Scafuro, A. C. 1997. *The Forensic Stage: Settling Disputes in Greco-Roman New Comedy*. Cambridge.

Sealey, R. 1983. "Athenian Courts for Homicide." *CP* 78: 275–96.

Sidwell, K. 1996. "Purification and Pollution in Aeschylus' *Eumenides*." *CQ* 46: 44–57.

Spence, I. G. 1993 *The Cavalry of Classical Greece*. Oxford.

Starkie, W. J. M. 1897. *The Wasps of Aristophanes*. London.

Stroud, R. S. 1994. "The Aiakeion and Tholos of Athens in POxy 2087." *Zeitschrift fur Papyrologie und Epigraphik* 103: 1–9.

1998. *The Athenian Grain-Tax Law of 374/3 B.C.* Athens.

Stumpf, G. 1987. "Zwei Gerichtsurteile aus Athen: *IG* II^2 1641 und 1646a." *Tyche* 2: 211–15.

Taylor, J. G. 1998. "Oinoe and the Painted Stoa: Ancient and Modern Misunderstandings?" *AJP* 119: 223–43.

Thompson, H. A. 1954. "Excavations in the Athenian Agora: 1953." *Hesperia* 23: 31–67.

Thompson, H. A. and R. E. Wycherley. 1972. *The Agora of Athens: The History, Shape and Uses of an Ancient City Center*. Princeton.

Thomson, A. 1898. *Euripides and the Attic Orators*. London.

Todd, S.C. 1993. *The Shape of Athenian Law*. Oxford.

Tilley, C. Y. 2010. *Interpreting Landscapes: Geologies, Topographies, Identities* Walnut Creek, CA.

Travlos, J. 1971. *Pictorial Dictionary of Ancient Athens*. London.

von Gall, H. 1979. "Das Zelt des Xerxes und seine Rolle als persicher Raumtyp in Griechenland." *Gymnasium* 86: 444–62.

Waddell, G. 1981. *Robert Mills's Courthouses and Jails*. Easley, SC.

Wallace, R. W. 2005. "Law, Attic Comedy, and the Regulation of Comic Speech." In *The Cambridge Companion to Ancient Greek Law*, ed. M. Gagarin and D. Cohen. Cambridge. 357–73.

Wiles, D. 1997. *Tragedy in Athens: Performance Space and Theatrical Meaning*. Cambridge.
Wohl, V. 2010. *Law's Cosmos: Juridical Discourse in Athenian Forensic Oratory*. Cambridge.
Worthington, I. 1992. *A Historical Commentary on Dinarchus. Rhetoric and Conspiracy in Later Fourth-Century Athens*. Ann Arbor.
Wycherley, R. E. 1953. "The Painted Stoa." *Phoenix* 7: 20–35.

INDEX

For EU product safety concerns, contact us at Calle de José Abascal, 56–1°,
28003 Madrid, Spain or eugpsr@cambridge.org.

www.ingramcontent.com/pod-product-compliance
Lightning Source LLC
LaVergne TN
LVHW011524200926
848484LV00001B/148
* 9 7 8 1 1 0 7 0 4 2 1 2 4 *